Gender Dimensions in Disaster Management

A Guide for South Asia

**Madhavi Malalgoda Ariyabandu
Maithree Wickramasinghe**

ITDG South Asia publication

*This publication was financially supported by the
Conflict and Humanitarian Affairs
Department of the UK Government's
Department for International Development (DFID)*

Practical Action Publishing Ltd
25 Albert Street, Rugby, CV21 2SD, Warwickshire, UK
www.practicalactionpublishing.org

© Intermediate Technology Publications 2003

First published 2003\Digitised 2013

ISBN 10: 1 85339 607 9
ISBN 13 Paperback: 9781853396076
ISBN Library Ebook: 9781780445465
Book DOI: http://dx.doi.org/10.3362/9781780445465

All rights reserved. No part of this publication may be reprinted or reproduced or utilized in any form or by any electronic, mechanical, or other means, now known or hereafter invented, including photocopying and recording, or in any information storage or retrieval system, without the written permission of the publishers.

A catalogue record for this book is available from the British Library.

The authors, contributors and/or editors have asserted their rights under the Copyright Designs and Patents Act 1988 to be identified as authors of their respective contributions.

Since 1974, Practical Action Publishing has published and disseminated books and information in support of international development work throughout the world. Practical Action Publishing is a trading name of Practical Action Publishing Ltd (Company Reg. No. 1159018), the wholly owned publishing company of Practical Action. Practical Action Publishing trades only in support of its parent charity objectives and any profits are covenanted back to Practical Action (Charity Reg. No. 247257, Group VAT Registration No. 880 9924 76).

Acknowledgements

Many individuals and organizations have contributed towards this publication.

First and foremost, we would like to thank the communities who shared their life experiences of living with disasters as a basis for this Guide.

We thank Sepali Kottegoda of the Women and Media Collective, Colombo for compiling the early drafts of the Practitioners' Guide, and for contributing to the discussions on the analytical framework of the Guide.

Thanks are extended to Bilquis Tahira, Consultant, Islamabad for compiling the section on 'Recommended Reading', and for her inputs to the section on the conceptual framework.

Special words of thanks are extended to the teams of ITDG Bangladesh, ITDG Nepal, Disaster Mitigation Institute, Ahamedabad, and Journalists Resource Centre, Islamabad, who were partners with us in coordinating the research case studies for the 'Livelihood Options for Disaster Risk Reduction' Project.

We are also indebted to the teams at the Journalists Resource Centre (JRC) and Rural Development Policy Institute, Islamabad, for coordinating the pre-test of the early draft, and for the continuous support extended throughout the printing process.

Ramona Miranda, Jayantha Gunasekara, Vishaka Hidellage, Susil Perera, and Louise Platt of ITDG South Asia, are thanked for their valuable contributions to the discussion sessions focusing on the structure of the book. The involvement of

Kusala Wettasinghe and Vindhya Tissera in the brainstorming discussions is much appreciated. A special note of thanks goes to Ramitha Wijetunga and Rohana Weragoda of ITDG South Asia, for preparing the statistics and proof reading. Sharon de Alwis, Sandya Wicramarachchi and Kay Goonathilake of ITDG South Asia are thanked for secretarial assistance.

A big thank you to Santosh Kumar of the World Bank office, New Delhi for writing the preface, despite a busy schedule.

We thank Sharni Jayawardena of Young Asia Television, Colombo, for editorial inputs.

Finally, we thank Conflict and Humanitarian Affairs Department (CHAD) of the Department For International Development (DFID) - UK for extending financial support for the research and for the publication.

Madhavi Malalgoda Ariyabandu
Maithree Wickramasinghe

Colombo
December 2003

About the authors

Madhavi Malalgoda Ariyabandu is a development researcher with special interest on political economy of development and disasters. She has designed and coordinated a number of research and training initiatives on livelihood options in disaster risk reduction and gender issues in disaster; and has analyzed disaster management policies with special reference to South Asia. She has considerable experience in interacting with the communities living with disasters in South Asia, based on which she has written a number of articles and papers focusing on gender issues. Madhavi authored the book **'Defeating Disasters: Ideas for Action'**, and co-edited the publications **'Seeing Disasters Differently-Visions and Suggestions'**, **'Disaster Communication, A Resource Kit for Media'**, and **'Disaster Risk Reduction in South Asia'**.

Madhavi is leading the Disaster Mitigation team at ITDG South Asia.

Maithree Wickramasinghe is a Senior Lecturer at the Department of English, University of Kelaniya, Sri Lanka, and a visiting lecturer on gender and women's studies at other educational institutions.

Her research work has explored a number of diverse issues including sexual harassment and violence against women, feminist critical theory and methodology, gender in organizations and workplaces, women and development etc. She is the author of **From Theory to Action - Women, Gender and Development** published by Friedrich Ebert Stiftung, as well as a number of other studies and articles on gender and women's issues.

ITDG is an international development agency with the vision "A world free of poverty and injustice in which technology is used to the benefit of all". It was begun by E.F.Schumacher, the economist and author of the widely-read book **Small is Beautiful**. ITDG has been in operation in Sri Lanka since 1989. ITDG South Asia currently works in the areas of energy, transport, manufacturing, agro processing, and disaster mitigation.

Duryog Nivaran is a network of individuals and organizations working in South Asia who are committed to promoting an alternative perspective on disasters and vulnerability as a basis for disaster mitigation in the region. The network's aim is to reduce the vulnerability of communities to disasters and conflicts by integrating the alternative perspective at conceptual, policy and implementation levels of disaster mitigation and development programmes in the South Asian region.

ITDG South Asia is a member organization of Duryog Nivaran.

Duryog Nivaran Secretariat
c/o ITDG South Asia
5, Lionel Edirisinghe Mawatha,
Kirulapone, Colombo 5,
Sri Lanka

Tel: +94-11-2829412
Fax: +94-11-2856188
Email: dn.net@itdg.slt.lk

About the book

This book aims to address the dearth of specific information on the subject of 'gender issues in disasters', particularly in the South Asian countries. The primary objective of compiling this Guide is to introduce the subject, and to raise awareness in policy/decision makers and the many thousands of development practitioners across South Asia, whose contribution is crucial for effective disaster management and sustainable development.

The book places the issue of gender in the context of development, and extends the discussion to show how gender and development concerns are reflected in the context of disasters.

The discussion of the book is based on the fundamental arguments that:

> The risk posed by natural hazards is a variable, and has direct implications on development in general, and livelihoods in particular.
>
> Disaster risk management is part of 'managing the livelihoods' for many millions of people in the sub-continent and elsewhere in the developing world.
>
> The gender concerns expressed and experienced in the development context are applicable in the context of disasters- with an added weight-due to the specific nature of vulnerabilities and capacities prevalent in all stages of disasters.
>
> The specific vulnerabilities and capacities of men and women, as well as the gender/social dynamics of disaster situations are often not obviously visible. Detailed livelihood analysis however, exposes these often subtle but vital considerations. If ignored or unattended, these concerns will impede development efforts from reaching their goals.

This Guide captures the experiences of ITDG, and the members of the Duryog Nivaran network, as they interacted closely with the communities living in situations of various natural hazards in South Asia during the last few years. While the major part of the evidence was generated through the case studies carried out in Bangladesh, India, Nepal, Pakistan, and Sri Lanka under the South Asia regional programme - 'Livelihood Options for Disaster Risk Reduction in South Asia' (implemented by ITDG South Asia), the book also draws on other documented evidence on the subject. It presents real-life examples, and case studies, which depict the subtle gender concerns and gender-based social dynamics prevalent in managing disasters, protecting daily-livelihoods, and in disaster/crisis situations.

In this Guide, ITDG and its network of partners in South Asia, present their firm conviction that incorporating disaster risk into development planning, and addressing gender considerations in all situations, is an absolute must to reach the goals of sustainable development and effective disaster risk reduction.

The guidelines proposed here aim to help address these concerns in planning and implementing development and disaster management programmes. The guidelines are presented under the two categories of 'Guidelines for Policy-makers' and 'Guidelines for Practitioners'. Of course, there are substantial linkages between the two sets of guidelines because they share the basic principles relating to interventions and arise from practical scenarios. While they stand independently of one another, cross-reference between the two sets will add to their completeness.

It must be cautioned, however, that the guidelines are not designed to be exhaustive; thus, policy-makers and practitioners must ensure that these recommendations are relevant/appropriate to the specificities of each situation.

Contents

Preface		**11**
1.	**Introduction**	**15**
1.1	Disasters – a growing problem	15
1.2	Seeing disasters differently	17
1.3	Understanding vulnerability to disaster	18
1.4	Linking disasters and development	21
1.5	Current "development" – a cause of disasters?	29
2.	**An alternative approach to disasters**	**31**
2.1	The Alternative Perspective	31
2.2	The issue of gender	34
2.3	Gender issues in disasters	42
3.	**Impact of disasters on women and men** *Realities for South Asia*	**51**
3.1	Differences in social and cultural impacts	51
3.2	Differences in economic impacts	60
3.3	Differences in psychological impacts	66
4.	**Recognizing community capacities**	**71**
4.1	Gender-based differences in coping with disasters	72
4.1.1	Gender-based differences in community preparedness for disaster	73
4.1.2	Coping as disaster strikes	75
4.1.3	Re-building after a disaster	77

5.	**Current practice** *The absence of gender sensitivity* *in disaster management*	**79**
6.	**Agendas for change** *Policy and practice*	**87**
7.	**Agenda for policy makers**	**91**
8.	**Guidelines for disaster management practitioners**	**111**
	Overall disaster preparedness	112
	Focus on women specific concerns	119
	Planning initial disaster responses/ emergency management	121
	Planning for rehabilitation/reconstruction	126
	Monitoring and Evaluation	128

Glossary of terms   **131**

Recommended reading   **155**

Preface

It is a pleasure to write a preface to a book addressing an issue that greatly concerns me - gender issues in disasters.

There are many myths about the whole concept of gender. In 'Gender Dimensions in Disaster Management: A Guide for South Asia', the authors have very nicely and lucidly tried to demystify the concept. Yes, gender is not another word for women; gender does not mean women and their problems; nor is it about mere biological differences between women and men. Gender is a social concept, and it is determined by social and cultural realities specific to each community or groups of people. As it is visible across the globe, social reality overplays the biological differences between men and women.

Why gender?

To begin with, the question may be asked as to why we should be concerned about the issues of gender at all. To respond in one sentence, development efforts are not designed to benefit 'only men' or 'only women'. Rather, officially, to benefit both - men and women. However, the real picture is that when it comes to implementation, a larger share of benefits and resources goes to men, and women continue to remain marginalized. 'Role stereotyping' guided by socially accepted norms largely excludes women from the mainstream of development planning. The execution of development programmes and their outcomes depend on the premise of 'inclusion' and 'exclusion', where men get included and women are left out. Indicators such as the Human

Development Index (HDI), Gender Development Index (GDI) and Gender Poverty Index (GPI) reflect this outcome. The UNDP Human Development Report draws attention to the persistence of severe gender disparities in human development. The central message of the report is that human development is not 'engendered', but 'endangered'. This is true for almost all the South Asian countries. The present book, arising from case studies carried out by ITDG and partner organizations (especially in South Asian countries), makes a commendable effort to look at the social/cultural realities and resulting differential vulnerabilities of men and women. Disaster management is a new 'sector' in development, which requires urgent attention. Unfortunately, this has been handled from a completely male-centered or male-focused perspective. A number of studies have testified to the dominance of men and masculine culture in this sector (Myers 1994; Morrow and Enarson 1996; Fordham and Ketteridge 1998).

Different vulnerabilities

This book puts forward convincing arguments that the disaster management approach should be revisited and reframed to make it more gender sensitive. Based on case studies conducted in the South Asia region, this book presents an in-depth analysis of the differential vulnerabilities of men and women in facing and managing disasters.

The Beijing Platform for Action, adopted at the Fourth World Conference on Women (1995), recognized the impact of environmental disasters on women, and proposed that women's responses to crisis be further investigated. Five years later, the Review and Appraisal of the Implementation of the Beijing Platform for Action (2000) identified natural disasters and epidemics as emerging issues that deserved greater attention. It was noted that the social and economic impact of natural disasters and epidemics remained relatively invisible as a policy issue. Their impact on the status of women, gender relations, and the achievement of gender equality has been almost completely ignored. In response to the findings of the Review and Appraisal, the twenty-third special session of the General Assembly entitled "Women 2000: Gender-equality, Development and Peace for the Twenty-first Century" acknowledged an increase in casualties and damages caused to women by natural disasters. It raised awareness, from a gender perspective, of the inefficiencies and inadequacies in existing approaches and intervention methods when responding to such emergency situations. The special session suggested that a gender perspective be incorporated into disaster prevention, mitigation and recovery strategies.

Identifying capacities

The Yokohama World Conference on Natural Disaster Reduction (1994) - a mid-term review of the International Decade for Natural Disaster Reduction, placed greater emphasis on the role of the social sciences in research, policy development and implementation, and emphasized the links between disaster reduction and sustainable development. It also recognized the need to stimulate community involvement and the empowerment of women at all stages of disaster management programmes as an integral part of reducing community vulnerability to natural disasters.

However, gender differences in disaster mitigation have been discussed primarily in the context of vulnerability or community involvement. Women's abilities to mitigate hazards and prevent disasters, and to cope with and recover from the effects of disasters have not sufficiently been taken into account, or developed.

Women are active throughout the disaster management cycle: in mitigation, prevention, preparedness, emergency response and recovery. Although their activities remain invisible and are undervalued, their importance cannot be ignored any more.

Gender sensitivity in disaster management can be ensured through a gender analysis of the situation at the very outset. This can be done with the help of tools such as the mapping of gender relations, time-use analysis of men/women, accounting for men's/women's access to and control over key resources, and their differential coping strategies/mechanisms/vulnerabilities/ capabilities. Such an analysis would help plan activities in a much more focused and realistic manner, which would not only maximize risk reduction, but also lead to an optimum utilization of resources, leading to the desired returns of socio-economic well-being.

This book has very effectively tried to provide pragmatic guidelines for incorporating gender analysis into disaster mitigation and preparedness planning, which I feel strongly to be crucial for development. I am confident the present piece of work will help policy planners, disaster managers, trainers and researchers gain new insights in mainstreaming the subject, and lead to further developments such as more effective disaster management plans and outcomes.

Dr. Santosh Kumar
Disaster Management Specialist
The World Bank
New Delhi
India

Introduction

1.1 Disasters – a growing problem

Disasters pose a serious threat to all aspects of development. Disasters result in death; and they cause physical, environmental and economic damage. Disasters challenge development by destroying social stability, and diverting already scarce resources to emergency responses. Huge amounts of money are spent by governments, businesses and communities to deal with the consequences of disasters. Global economic costs related to disaster events average around US$ 880 billion per year.

Nearly 90 percent of natural disasters and 95 percent of disaster-related deaths worldwide occur in developing countries. It is estimated that by the year 2025, 80 percent of the world's population will live in developing countries, and up to 60 percent of them will be highly vulnerable to floods, severe storms and earthquakes.[1]

[1] Moin F, 'Disasters and Development' in *Disaster Mitigation – Experiences and Reflections*, Sahni P, Dhameja A, Medury U., eds., New Delhi, 2001

[2] Here, the definition of disasters includes natural as well as man-made disasters. International Federation of Red Cross and Red Crescent Societies, World Disaster Report 2002, Geneva 2002

[3] ibid

> During the last decade (1992 - 2001) disasters have claimed 96,285 lives in the South Asian sub continent.[2]

In 2001:

- 56 percent of people killed by disasters across the world lived in South Asia.
- 36,651,662 people were affected by disasters in India.
- 1,315,211 people were affected in Pakistan.
- The total number of people who got affected in Bangladesh, Nepal and Sri Lanka was 729,033, 21,026 and 1,000,200 respectively.[3]

What causes alarm is the rising trend in losses. Worldwide losses from disasters during the 1990s are almost 3 times greater than those recorded from 1981 to 1989.[4]

Because of its geographical location, the South Asian sub-continent is exposed to a variety of hazards such as floods, drought, cyclones and earthquakes. While these and other natural hazards are not novel experiences for the people living in the region, the disasters leading from these hazards are increasing. All countries in the globe are affected by some type of natural hazard; but the magnitude of the hazard, how it is managed, and consequently, its impact, vary widely.

During the past decade alone, natural hazards in South Asia resulted in major disasters with thousands of deaths and massive destruction: namely, a cyclone in Orissa in India (1999); earthquakes in Latur (1996), and Gujarat (2001) in India; a cyclone in Bangladesh (1991); and flash floods in Nepal (1993). Although these were disasters that received extensive reportage, there were hundreds of frequent and smaller disasters that did not receive publicity.

[4] International Decade for Natural Disaster Reduction (IDNDR), Final Report of the Scientific and Technical Committee, Geneva, 1999

[5] International Federation of Red Cross and Red Crescent Societies, World Disaster Report 2002, Geneva, 2002 p 200

Table 1: Impact of Disasters in South Asia for the 20-year period 1982 -2001.[5]

Country	Annual average number of people killed	Annual average number of people affected
Bangladesh	8,754	15,897,987
India	5,391	56,116,660
Pakistan	542	1,437,785
Nepal	285	92,497
Sri Lanka	65	727,182
20 year total	300,740	1,485,442,220

From 1992 to 2001 there were 622,363 deaths from disasters and 2,001,519,000 people were affected throughout the world. The total damage was an estimated US$ 694,424 million, globally.[6]

[6] Op. cit. pp 196, 202

1.2 Seeing disasters differently

Disasters are usually seen as sudden events that result in death and destruction; and which require immediate, emergency relief. Natural hazards are usually cited as the cause of disasters and are sometimes viewed as 'acts of the gods' and divine punishment. To date, disasters have largely been the concern of environmental and geological scientists who study natural hazards, and the humanitarian institutions that have to deal with the human costs.

The term 'natural hazard' is often confused with the term 'natural disaster'. In order to understand disasters better, the relationship between these two key concepts and other related ideas must to be explored.

A hazard can be defined as a phenomenon that has the potential to cause injury to life, livelihoods and habitats. High winds, floods, landslides, droughts and earthquakes are all natural hazards. If unmanaged, that is, if natural hazards and their potential consequences are not prepared for, they become **disasters** that take lives and damage livelihoods.

Vulnerability is a set of conditions that affect the ability of countries, communities and individuals to prevent, mitigate, prepare for and respond to hazards. High levels of vulnerability increase the likelihood that natural hazards will be unmanaged and result in disaster.

Although most natural hazards have the potential to be very destructive, they do not always turn into disasters. It is the combination of a natural hazard and vulnerability that leads to a disaster.

"A disaster occurs when a significant number of vulnerable people experience a hazard and suffer severe damage and/or disruption of their livelihood system in such a way that recovery is unlikely without external aid."[7]

[7] Blaikie P, Canon T, Davis I, Wisner B, At Risk natural hazards, people's vulnerability, and disasters, London, Routledge, 1994. p 21

[8] Maskrey A, Disaster Mitigation: A Community Based Approach, Development Guidelines, No 3, OXFAM, UK, 1989

[9] Ariyabandu MM, Defeating Disasters: Ideas for Action, ITDG, Colombo, 1999, p 8

As disasters continue to affect more and more people, it becomes increasingly clear that the social, economic and political forces of development processes contribute to disasters. Aside from the sudden hazards that arise from nature, people may also be made vulnerable due to the changes wrought by the unequal distribution of resources, social and political processes that perpetuate inequality, global macro-economic forces, and uncoordinated, ill-planned development. Consequently, there is growing recognition of a need to re-examine how we view and act towards disasters.

1.3 Understanding vulnerability to disaster

Some groups are more vulnerable than others. Vulnerability is not just poverty, but the poor tend to be the most vulnerable due to their lack of choices. The influences of both poverty and development processes on people's vulnerability to disasters are now well established.[8] Class, caste, ethnicity, gender, disability and age are other factors affecting people's vulnerability.[9]

Because vulnerability plays such an important part in why natural hazards become human disasters, it is worth spending some time to examine the characteristics of vulnerability.

Conditions of vulnerability are a combination of factors that include poor living conditions, lack of power, exposure to risk, and the lack of capacity to cope with shocks and adverse situations. It can be helpful to think of vulnerabilities in the following way:

- **Physical vulnerabilities** are the hazard-prone locations of settlement, insecure and risky sources of livelihood, lack of access to basic production resources (such as land, farm inputs, and capital), lack of knowledge and information, and lack of access to basic services.

- **Social vulnerabilities** are reflected in the lack of institutional support structures and leadership, weak family and kinship relations, divisions and conflicts within communities, and the absence of decision-making powers.

- **Attitudinal vulnerabilities** are seen in dependency, resistance towards change, and other negative beliefs.[10]

As noted earlier, poverty does not equal vulnerability, but being poor makes people more vulnerable to disasters because poor people lack the resources (physical, social, and knowledge-based), to prepare for and respond to such threats and shocks as natural hazards.[11]

Poor people often get locked in a cycle of vulnerability. Because they are poor, they become vulnerable. Because they are vulnerable, they are at great risk in the face of a natural hazard - leading to disaster. Because they suffer greater losses from a disaster, they become even poorer, more vulnerable, and are at an even greater risk of another disaster.

[10] Anderson MB, Woodrow PJ, Rising from the Ashes: Development Strategies in Times of Disasters", Westview Press and UNESCO, (1989), pp 13-14

[11] Cuny, FC, Disasters and Development, INTERTECT Press, Texas, USA, 1994

The spiral of poverty, vulnerability and disaster risk

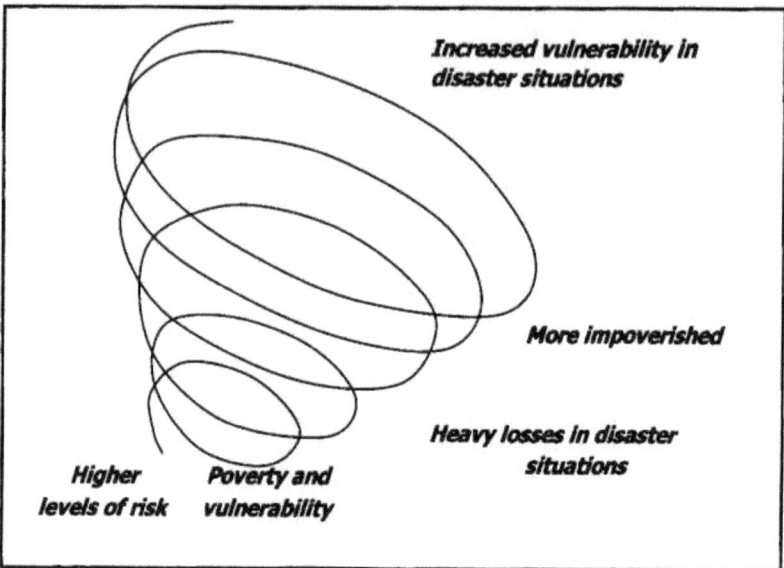

Figure 1

There is growing awareness that:

- Regardless of location, there is a direct link between poverty and powerlessness, and the severity of the impact of a disaster.

- Socially and economically vulnerable groups in any society are worst affected by natural hazards.

- There is a cycle of vulnerability to disasters – vulnerable groups are made even more vulnerable by disasters. This reduces their ability to pick up the pieces after a disaster and makes them further vulnerable in the face of another disaster.

1.4 Linking disasters and development

There are a number of approaches and tools for examining the links between vulnerability and disasters (See boxes below). These are important steps towards identifying and understanding the multiple causes of disasters, and can be the basis of effective disaster management policy and practice. They illustrate how reducing vulnerability is a key element in disaster management efforts, and argue for the need to integrate it into overall development policy.

A simple tool proposed to understand and explain the causes of disaster, and show 'how disasters occur when natural hazards affect vulnerable people' is the **Disaster Pressure and Release Model (PAR)**.[12]

Disaster Pressure and Release Model (PAR)

"The basis for the Pressure and Release (PAR) idea is that a disaster is the intersection of two opposing forces: those processes generating vulnerability on the one side, and physical exposure to hazard on the other".[13] To relieve (release) the pressure, vulnerability has to be reduced.

By analysing the social processes that increase people's vulnerability to natural hazards and risk of disaster, PAR also helps to show that the causes of a disaster may not be immediately obvious or visible.

[12] See Blaikie P, Cannon T, Davis I, Wisner B, At Risk: natural hazards, people's vulnerability, and disasters, London, Routledge, 1994, pp 21-45, for a full explanation of the Disaster Pressure and Release Model.

[13] Op. cit. p 22

Disaster Pressure Model

The progression of vulnerability

1	2	3	
ROOT CAUSES	**DYNAMIC PRESSURES**	**UNSAFE CONDITIONS**	**HAZARDS**
Limited access to • Power • Structures • Resources **Ideologies** • Political systems • Economic systems	**Lack of** • Local institutions • Training • Appropriate skills • Local investments • Local markets • Press freedom • Ethical standards in public life **Macro forces** • Rapid population growth • Rapid urbanization • Arms expenditure • Debt repayment schedules • Deforestation • Decline in soil productivity	**Fragile physical environment** • Dangerous locations • Unprotected building and infrastructure **Fragile local economy** • Livelihoods at risk • Low income levels **Vulnerable society** • Special groups at risk • Lack of local institutions **Public actions** • Lack of disaster preparedness • Prevalence of endemic disease	Earthquake High winds (cyclones/hurricanes/typhoon) Flooding Volcanic eruption Drought Virus and pests

RISK = Hazard + Vulnerability

R = H + V

Figure 2

**Box 1
Disaster
Pressure and
Release
Model (PAR)**

The diagram illustrates three levels of 'social factors' that generate vulnerability, thereby increasing people's risk of disaster from hazards.

1. **Root or underlying causes**. The most important root causes of vulnerability are the economic, demographic and political processes that affect the allocation and distribution of resources between different groups of people. Root causes reflect the distribution of power in society - including gender relations.

2. **Dynamic pressures.** These are processes and activities that impact on the root causes leading to particular forms of vulnerability. Rapid population growth and urbanization, loans and debt re-payment, currency devaluation leading to rise of prices in basic needs and services, continuing deterioration of land due to erosion and deforestation, and growing demands on land continue to apply pressure on people living in the margins - pushing them towards unsafe conditions.

3. **Unsafe conditions.** Usually, these conditions are highly visible forms of vulnerability. Unsafe conditions can be categorised into:
    - Fragile physical environments
    - Fragile local economies
    - Vulnerable societies/groups
    - Public/political actions

Poor people are often pushed to settle in unproductive and barren lands, in drought-prone areas, in fragile flood plains, in mountain slopes subject to erosion and landslides. Poor economic conditions deny people opportunities of education, access to information, markets and health services. This results in fewer options for securing a livelihood, often forcing people into insecure, poorly paid, hard, and at times,

Disaster Pressure and Release Model (PAR)

dangerous ways of supporting themselves. Daily wage labour, load carrying, tenant cultivation, dependency on landlords, moneylenders and middlemen are common features in the livelihood systems of poor people. Lack of savings and insurance, little or no control over resources like land, capital and information, lack of influence and visibility in institutions and processes that affect their lives reflect the fragile nature of poor people's livelihood systems. The outcome is an increase in the number of people who are vulnerable to shocks like natural hazards and who are, therefore, at risk to disasters.

The Sustainable Livelihoods Framework (SLF)

The Sustainable Livelihoods Framework (SLF) presented in Box 2 illustrates how vulnerability and risk of disaster are part of people's livelihoods. The SLF presents an analysis of people's livelihood strategies as well as the major influences that impact on sustaining livelihoods.

Box 2 Sustainable Livelihoods Framework

[14] See Department for International Development (DFID) Guidance Sheets (1999-2000) on Sustainable Livelihoods Framework, www.livelihoods.org

Sustainable Livelihoods Framework – accounting for disasters and vulnerability in development

There are different Sustainable Livelihoods frameworks – this box introduces the framework popularized by the Department for International Development (DFID) UK.[14]

In the Sustainable Livelihoods Framework (SLF), a livelihood is defined as " ... the capabilities, assets (including both material and social resources) and activities required for a means of living. A livelihood is sustainable

Sustainable Livelihoods Framework

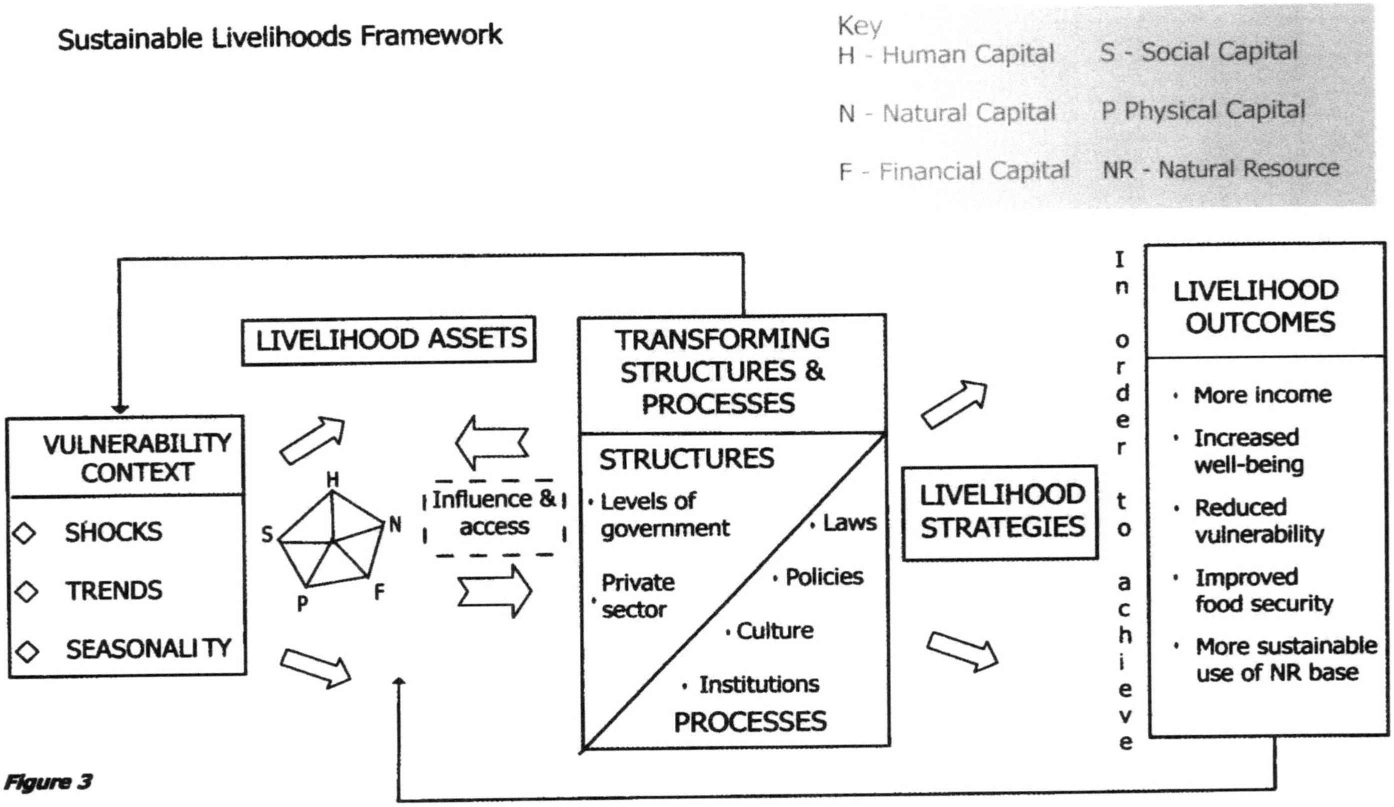

Key
H - Human Capital S - Social Capital
N - Natural Capital P Physical Capital
F - Financial Capital NR - Natural Resource

VULNERABILITY CONTEXT
- ◇ SHOCKS
- ◇ TRENDS
- ◇ SEASONALITY

LIVELIHOOD ASSETS

Influence & access

TRANSFORMING STRUCTURES & PROCESSES

STRUCTURES
- Levels of government
- Private sector

- Laws
- Policies
- Culture
- Institutions

PROCESSES

LIVELIHOOD STRATEGIES

In order to achieve

LIVELIHOOD OUTCOMES
- More income
- Increased well-being
- Reduced vulnerability
- Improved food security
- More sustainable use of NR base

Figure 3

The Sustainable Livelihoods Framework (SLF)
Box 2 continued

when it can cope with and recover from stresses and shocks and maintain or enhance its capabilities and assets both now and in the future, while not undermining the natural resource base".[15]

SLF views people as operating in a *vulnerability context*. The vulnerability context " ... frames the external environment in which people exist and is responsible for many of the hardships faced by the world's poorest people. The factors that make up the vulnerability context are important because they have a direct impact upon people's assets and the livelihood options that are open to them".[16]

There are three categories of vulnerability:
- **Trends** are the demographic, economic, technological and political trends.

- **Shocks** include health shocks like epidemics, natural disasters, conflict, and shocks to crops/livestock.

- **Seasonality** are shifts in prices, production, food availability, employment opportunities and health.[17]

Strengths and capacities of people are expressed in terms of their livelihood assets (represented by the pentagon in figure 3). Livelihood assets ' ... are created and destroyed as a result of the trends, shocks and seasonality of the vulnerability context'[18].

Transforming structures are 'the institutions, organisations and legislation that shape livelihoods ... their importance cannot be over

[15] DFID, (Department for International Development), Sustainable Livelihoods Guidance Sheets, London, 1999/2000

[16] Twigg J., Sustainable Livelihoods and Vulnerability to Disasters, *Disaster Management Working Paper 2/2001*, Benfield Greig Hazard Research Centre, University College of London, http://www.bghrc.com

[17] Op. cit. pp 10

[18] Op. cit. pp 11

The Sustainable Livelihoods Framework (SLF)

emphasized ... they can reduce or worsen the impact of external shocks on vulnerable people'. These transforming structures and processes determine people's access to livelihood assets, options, decision-making, as well as the power relationships that affect terms of exchange, and the benefits gained by people from livelihood strategies.

"Operating within the vulnerability context, using their livelihood assets and under the considerable influence of transforming structures and processes, poor people choose and implement livelihood strategies. These are often complex and may change in response to the external context. The SLF seeks to understand the many factors influencing people's choice of livelihood strategy, and then to reinforce the positive aspects (factors that promote choice and flexibility) and mitigate the constraints ... Placing vulnerability and external shocks at the heart of livelihoods analysis is a big step forward from much conventional development thinking ... Overall, the SL framework is a good model for viewing livelihoods in all their aspects, and in setting risk reduction and hazard vulnerability in the wider vulnerability and livelihoods context".[19]

---

[19] Op. cit. pp 11 - 13

The Sustainable Livelihoods Framework (SLF)

> **The SLF helps to illustrate how:**
> - **Coping with natural hazards and risk management is an integral part of sustaining and managing livelihoods.**
> - **Levels of livelihood assets of communities/individuals reflect their vulnerability.**
> - **Vulnerabilities and degree of risk to disaster differ for men and women.**
> - **There is a gendered division of labour in how communities and individuals cope with disaster (before, during and after disasters).**
> - **Women and men's capacities for coping with risk and disaster are different and related to the gendered division of labour.**

Like the PAR Model in Box 1, the SLF explores the links between wider or macro processes and structures, vulnerability, and people's livelihoods. The difference between PAR and Sustainable Livelihood approaches is that the PAR takes disasters and vulnerability to hazards as the starting point of analysis. SLF places livelihoods at the centre of analysis and identifies a whole range of vulnerabilities, including disaster risk from natural hazards, as part of a context that shapes people's livelihood strategies and outcomes.

The ultimate development goal of the SLF is sustainable livelihoods, and ensuring livelihood security is the way to achieve this goal. Reducing vulnerability and risk through a dual process of reducing constraints and building up strengths/ capacities is key to ensuring sustainable livelihoods and breaking the cycle of vulnerability and disaster.

Hitherto, most development approaches, such as 'Poverty Alleviation', the 'Basic Needs Approach', the 'Trickle Down Theory' or the 'Integrated Rural Development Strategy' have not included disaster risk and disaster mitigation either in their analysis of problems or in developing strategies. Ignoring disasters in development policy and practice has meant that development strategies have often failed. This is due to the fact that disasters are often a recurring phenomenon in some areas and result in development efforts being constantly defeated. Consequently, the success and sustainability of development can only be assured if the risk of disasters is accounted for in development approaches.

Sustainable Livelihoods Framework is part of a more recent approach that is still being developed; but is gaining popularity with development organizations. SLF is also seen as an opportunity to raise the agenda to account for disasters and vulnerability in development policy and practice.

1.5 Current "development" – a cause of disasters?

The last century was one of rapid growth and development in many parts of the world. But industrialization and the advance of science and technology have also had costs. The process of development has been literally fuelled by natural resources (land, minerals, water, forests, etc.) and has generated consumption patterns and social expectations that will continue this trend.

Sadly, growth and development are not balanced or equitable. Natural resources are used without

due thought of replacement, leading to resource degradation and depletion. The benefits from growth and development are not shared equally between countries and communities.

The world today is in a vicious cycle, where natural hazards and socio-economic development processes are combining to create new and greater conditions of vulnerability. As vulnerability increases, so do disasters. While disasters destroy development efforts, development itself creates more and more conditions, which result in disasters.

The consequences of ill-planned, uncoordinated and unsustainable development can be seen in South Asia, where the risk of disaster is increased by disagreements and conflicts between countries over the ownership and control of natural resources. Water is one example, where tensions in the region have increased people's vulnerability to drought, floods, landslides and other water-related natural hazards.

Nor do disasters recognize political borders. Investment and resource use in one country can increase or decrease the risk of disaster in another. For example, flooding in Bangladesh can be directly linked to watershed management in India.

As noted earlier, it is vital that we recognise that disasters do not arise only from natural hazards (like droughts, floods, landslides etc.) but that disasters also arise due to various political social and economic forces of development. Consequently, there are both 'socio-economic' as well as 'natural' sides to a disaster.

An Alternative Approach to disasters

Current disaster management practices in South Asia exhibit a combination of institutional structures, policies and programmes that focus on emergency responses and relief. This is the **Dominant Approach** - responses to disasters are usually unsystematic, and come from highly centralized, top-down, inflexible bureaucracies[1]. The communities at risk are rarely involved in decision-making. Although governments usually allocate funds for emergencies, these monies are not employed for disaster prevention and preparation; rather, they are withheld from use until a disaster actually strikes. The result is that an increasing percentage of national budgets and non-government aid is only spent on emergency response, thereby diverting resources away from development.

[1] Ariyabandu MM, Defeating Disasters, Ideas for Action, ITDG, Duryog Nivaran, Colombo, 1999

[2] Op. cit. p 19

2.1 The Alternative Perspective[2]

The Alternative Perspective advocated by Duryog Nivaran looks at disasters as part and parcel of the 'normal' development of societies - as unresolved problems of development. The line of reasoning here is that there is a strong linkage between a disaster and the conditions in society during 'normal' times.

> The Alternative Perspective looks at the underlying reasons as to why certain sections of society are more vulnerable to disasters than others. It argues that these links need to be understood, if we are to identify the causes and effects of disasters, and if we are to plan how to deal with them.

The Alternative Perspective proposes ways of mitigating disasters that would include all aspects

[3] Op. cit. p 20

of risk reduction, disaster preparedness, immediate relief, rehabilitation and long-term reconstruction to replace the Dominant Approach (which deals with disasters only after they happen). The Alternative Perspective recognizes that communities have an important role to play in disaster mitigation, and that community-based approaches should be used in these interventions. This involves taking needs-based approaches and strengthening the capacities of communities and individuals to plan/prepare for disasters and thereby reduce the levels of risks.

In the table that follows, 'Duryog Nivaran', the South Asian network for disaster mitigation, makes a comparative analysis of the dominant and alternative paradigms.

Dominant Perspective[3]	Alternative Perspective
Disasters/conflicts are viewed as isolated events.	Disasters/conflicts are seen as part of the normal process of development.
Linkages with conditions in society during normal times are not always analysed.	Analysing linkages with society during normal times is fundamental to understanding disasters/conflicts.
Technical/law and order solutions are dominant.	Emphasis on solutions that change relationships/structures in society. The objective is to reduce people's vulnerability and strengthen their capacity.
Centralized institutions dominate in intervention strategies and there is less participation of people, who are treated as 'victims'.	Decentralized institutions dominate in intervention strategies. Participation of people paramount in intervention strategies; people treated as 'partners' in development.

Dominant Perspective	Alternative Perspective
Implementing agencies less accountable, and their	Accountability and transparency
Interventions are made after the	The fundamental aim is the
The objective of intervention is to return to the situation before	Disasters/conflicts viewed as opportunities for social trans-

The International Decade for Natural Disaster Reduction (IDNDR) which concluded in 1999, states in its Geneva Mandate[4] on Disaster Reduction:

'We shall adopt and implement policy measures at the international, regional, sub-regional, national and local levels aimed at reducing the vulnerability of our societies to both natural and technological hazards through proactive rather than reactive approaches. These measures shall have as main objectives the establishment of hazard resilient communities and the protection of people from the threat of disasters.'

IDNDR calls upon national governments to demonstrate a policy commitment to reduce their vulnerability through declarations, legislation, policy decisions, and actions at the highest level. Accordingly, national governments in South Asian countries have accepted IDNDR's concepts and approaches in principle[5]. At the same time, international organizations such as UNDP[6], World Bank[7], and IFRC[8] have also taken them on board.

[4] International Decade for Natural Disaster Reduction (IDNDR), International Programme Forum, 'The Geneva Mandate on Disaster Reduction', Geneva, July 1999

[5] Bhatti A, Ariyabandu MM, *Disaster Communication, A Resource Kit for Media*, pp 36-37, Journalists Resource Centre, Islamabad, ITDG South Asia, Colombo, 2002

[6] http://www.undp.org/erd/disred/index.htm

[8] International Federation of Red Cross and Red Crescent Societies, *World Disaster Report 2002 Focus on Reducing Risk*, Geneva, 2002

[7] http://216.239.33.100/search?q=cache: http://wbln0018.worldbank.org/ html/smallstates.nsf/(SmallStateslookup1)/VulnerabilityVolatility%3FOpenDocument+disasters

2.2 The issue of gender

The dominant approach to disasters does not usually recognize or address the different vulnerabilities of women and men to disasters. There is comparatively little understanding of the 'gendered' aspects of risk and vulnerability to disasters not only in South Asia, but also worldwide.

The importance of development policy and practices that are sensitive to gender concerns is increasingly acknowledged, especially by the Alternative Approach. The arguments for mainstreaming gender perspectives into development arise from the failure to engage women in development processes as equal partners to men (either as decision-makers or beneficiaries).

This has meant that the powers and benefits of development have been confined to men, since the dominant development approach only targeted men; and saw 'heads of households', 'farmers', 'breadwinners' as men. Women were merely seen as 'housewives', 'secondary earners' and 'mothers' within the context of the family/household unit, and, if at all, only addressed in these roles.

Thus, for decades, the dominant approach to development has marginalized women – despite the fact that they constitute more than half the world's population.

Some of the main arguments for making development gender sensitive are:

- The fact that worldwide women are poorer than men.
- The different and unequal ownership, access, and control that women and men have over resources.

The dominant development approach saw 'heads of households', 'farmers', 'breadwinners' as men. Women were merely seen as 'housewives', 'secondary earners' and 'mothers' within the context of the family/household unit.

- A gendered division of labour that restricts the participation of women and men to certain activities and results in different needs, capabilities and capacities.
- The low status often given to women, and the belittling of the activities traditionally engaged in by women. For example, lower payment for 'women's work'.
- The lack of recognition (invisibility), and the devaluing of the economic and social contributions made by women.

The basic differences between men and women, as well as the resulting differences in their status, interests, and needs with regard to development can be understood with reference to the concept of gender.

Box 1

Understanding/defining gender

Gender refers to the way members of the two sexes are perceived, evaluated and expected to behave.

"Gender then refers to a whole set of expectations held as to the likely behaviour, characteristics, and aptitudes men and women will have. It refers to the social meanings given to being a man or women in a given society". Kate Young[9]

A **Gender Identity**[10] constitutes of the following:
- **a person's given/adopted gender roles responsibilities** (see Table 1)
- **the characteristics and conduct given for each sex**
- **the appearance and dress codes that are expected from men and women**
- **the professions that are assigned to men and women, and**
- **the sexual orientation/preferences of men and women.**

[9] Young K, Towards the theory of a Social Relation of Gender; Institute of Development Studies, Sussex, UK, 1988

[10] Wickramasinghe M, *Gender Identity and Gender Relations in Gender Resource Book for Teachers* Mendis SK ed, CENWOR, Colombo, 2002.

The norms governing the gender identities of men and women are not identical. In fact, the gender identities of a man and a woman may differ greatly on account of time and place and according to the accepted models of gender found in different religious, cultural and class ideologies and social structures. Deviations from these allocated gender standards (such as failure to subscribe to gendered behaviour patterns or accept gendered responsibilities) may result in social ridicule/condemnation/ostracization – especially for women.

Table 1 sketches what are considered to be the typical gendered roles and responsibilities of men and women (note that there may be cultural differences in gender roles/responsibilities in some societies).

Table 1

A Gender Roles / Responsibilities Framework[11]	
Women	**Men**
Women's productive roles and responsibilities	**Men's productive roles and responsibilities**
These include women's roles and responsibilities that give economic remuneration whether for manual labour; professional labour; subsistence activities; part-time work, or casual labour.	These involve men's roles and responsibilities that give economic remuneration whether for manual labour; professional labour; subsidiary activities; part-time work; casual labour, etc. **(Men are principally identified in relation to these roles and responsibilities).**
Reproductive roles and responsibilities	**Family roles and responsibilities**
These include women's roles and responsibilities within the household and the family: inclusive of bearing, nurturing, rearing children; cooking; cleaning the house and yard; washing and laundering clothes; fetching water/fuel-wood; marketing; caring for the sick and the elderly, etc. These roles may be expanded to include agricultural work in the homestead, work relating to livestock within the household, etc., which are not given economic value. **(Women are principally identified in relation to these roles and responsibilities).**	These are the occasions and the degree to which men are involved in household/family maintenance. Depending on the many variables of community, geography, age, and time-period, men may contribute in the provision of travelling; protection to the family; in household tasks, etc.

[11] From Wickramasinghe M, From Theory to Action-Women Gender and Development, Friedrich Ebert Stiftung, 2000 (based on the Harvard Framework, A Casebook: Gender Roles in Development Projects, Overholt C, Anderson MB, Cloud K, Austin JE (eds), West Hartford, CN. Kumarian Press, 1985).

A Gender Roles / Responsibilities Framework

Women	Men
Community roles and responsibilities	**Public roles and responsibilities**
Includes women's roles and responsibilities in the community: maintaining kinship relations; religious activities; social interactions and ceremonies (births/marriages/deaths); communal sharing and caring activities; community survival strategies; etc. **(These are done voluntarily and do not provide economic returns and are usually linked to their reproductive functions.)** However, in certain cultures, these activities become the province of men.	Involves men's public roles and responsibilities: their visibility in the public and powerful spheres - of politics; in decision-making bodies; in status-building activities; in international forums, etc. **(These give men power/ visibility/social or official status/higher financial rewards.)**

Women and men may also have certain roles and responsibilities with regard to natural resources and the environment:

Women	Men
These include women's roles and responsibilities as agricultural workers, home gardeners, conservationists, consumers, relief workers, providers of home remedies and indigenous medicines	These involve men's roles and responsibilities as consumers, weather watchdogs, relief workers, etc.

Gender Needs
The different needs which arise due to the differing gender identities of men and women.

Practical Gender Needs refer to those needs that arise from women's gender roles and responsibilities such as food, water, fuel, etc., and are usually related to immediately perceived basic needs.

Strategic Gender Needs refer to those needs that arise due to women's subordinate positions in society and include such needs as equal wages, right to live free from gender-based violence, legal rights etc., and are not always visible.

Gender Relations[12]
Gender relations refers to the actual and perceived network of relations that occur between men and women. These involve daily life experiences, as well as notions of gender relations, which emanate from the media, religions, history, culture, etc. Usually gender relations are unequal, because men have power and women do not. Women and men conform to what is accepted of them.

In general, women are more disadvantaged than men. Though men may face severe obstacles in life due to their ethnic, political, or socio-economic backgrounds, men do not usually face gender discrimination. **Not only are women subjected to these other forms of discrimination, they are also discriminated within their families, kinship structures, and communities due to accepted norms and images of gender identities.**

Vast numbers of women are economically underprivileged as opposed to men; and are, unlike men, targets of unequal cultural practices.

[12] Wickramasinghe M, *Gender Identity and Gender Relations* in *Gender Resource Book for Teachers* Mendis SK ed, CENWOR, Colombo, 2002

They are disenfranchised vis a vis politics and legal systems due to various gender-unequal ideologies and hierarchies.

One reason for this is that women are made to shoulder many responsibilities and negotiate multiple roles in life. Often, women end up working longer and harder than men, due to this very reason. Aside from their paid work to support families (women are often paid less than men for equal amounts of work; or wedged in underpaid occupations), the reproductive roles of women involve them, day after day, in recurring, arduous, and often exhausting tasks of the household; and in satisfying the needs of family members including those of children, the elderly, and the sick. These responsibilities may confine women to the domestic sphere or immediate home environs.

This can be seen to result in a gender division of labour; which places men in the more public and powerful areas of life that offer official/social positions and economic rewards. The ceaseless labour of women within the household, however, does not offer payment, and results in making women permanently poor. Often women's extensive labour (within economic/reproductive/ community roles) and lifestyles are largely invisible, and their differing needs unrecognized. In comparison, men's roles and responsibilities are publicly in evidence and better valued.

Moreover, unlike in the case of men, women are greatly affected by fixed and deep-rooted gender roles and trait stereotyping. For instance, social norms may hold women responsible for the lifelong nurture of the family. Culture may demand that women are placed in seclusion and restrict their mobility. Lesser valuation of women within the family may result in women eating less. Patriarchal ideologies may suppress women's voices through wife-beating. Often, women are viewed as being physically or intellectually less capable than men,

Often women's extensive labour (within economic/ reproductive/ community roles) and lifestyles are largely invisible, and their differing needs unrecognized.

and hence, their views are either not solicited, or disregarded in development planning.[13]

Unlike in the case of men, these powerful and unequal gender ideologies may obstruct women's access to material and other resources - both within the household and in society at large.

[13] Kottegoda S, 'A study of gender aspects of Communities living with drought and landslides in Sri Lanka', ITDG South Asia, Colombo, 2001

All in all, women, and especially poor women, are thus rendered vulnerable even before disaster strikes. Therefore, women's position can only be further exacerbated in the face of a crisis.

GENDER FACTORS INCREASING RISK FOR GIRLS AND WOMEN

- Childbirth- and pregnancy-related health limitations
- Longer life span and increased mobility limitations, chronic illness, disabilities
- Limited reproductive control
- Greater risk of domestic and sexual violence
- More likely to be economically dependent
- Less access to credit
- Fewer land rights
- Less control over labour
- More often employed as part-time, "flexible" workers, and in free trade zones
- More responsibility for dependents
- More dependent on child care centres, schools, clinics, and other public services
- Less access to transportation
- Higher illiteracy rates, lower levels of schooling and training
- More dependent on water, fuel wood, crops and other natural resources
- Less free time and personal autonomy
- More often socially isolated
- Less decision-making power in homes and political institutions

[14] Source: Enarson E, Lourdes M, Betty Hern M. Working with women at risk, practical guidelines for assisting local disaster risk, April 2002, p 6

- Subject to "intersecting vulnerabilities" (e.g. as impoverished women raising families in substandard housing; under-employed disabled women subject to sexual violence; frail older women who are illiterate etc.)
- Low representation in emergency management organizations and other professions
- Less knowledge of how to access emergency assistance or capacity to do so[14]

2.3 Gender issues in disasters

The two sexes experience disasters differently due to their gendered social construction. Consequently, disasters also impact differently on men and women. At the same time, men and women's roles in the disaster management process are different.

Close analysis of disaster impact shows that the vulnerability of men and women to disasters, their capacities, and the options available to them differ in character and scale according to their gender.

A woman may have access to indigenous knowledge that portends disasters, while a man may rely on weather bulletins. A woman's daily workload may increase during a disaster, whereas a man's work may actually decrease. A man may be able to migrate in search of employment during a crisis period, while a woman may be left with the responsibility of the family. Unlike a man, a woman may not receive emergency relief to her hand, as cultural customs may not allow women direct contact with male relief workers.

Analysis further indicates that although women are often more vulnerable to disasters than men (owing to conventional gender responsibilities and relations), they are not just helpless victims as often represented. Women have valuable knowledge and experience in coping with disasters. Yet, these strengths and capabilities of women are often ignored in policy decisions, and in mitigation, thereby, allowing these valuable resources to go waste, and sometimes creating dependency situations.

Thus, ignorance of gender differences has led to insensitive and ineffective relief operations that largely bypass women's needs and their potential to assist in mitigation and relief work.

Women and men perform distinct disaster-preparedness activities. In many communities, women take an active part in community disaster initiatives - both in roles of leadership and at grassroots – outnumbering men. Yet, in larger, more formal emergency planning organizations, women are scarcely represented, and markedly absent from decision-making positions. This appears to be true for both the developed and developing world.[15]

Not being sensitive to gender issues in development planning and disaster mitigation means that interventions are often only targeted at men. Inaccurate gender assumptions of policy-makers and practitioners may not only deny women benefits, but also serve to worsen the situation for women. Sensitivity to gender implications is vital, in order to empower a community to successfully move on and move up from the abyss of a disaster.

In many communities, women take an active part in community disaster initiatives. Yet, in larger, more formal emergency planning organizations, women are scarcely represented, and markedly absent from decision-making positions.

[15] Enarson E, Hearn Morrow B, *The Gendered Terrain of Disaster, Through Women's Eyes*, Praeger, London, 1998

Box 2 — **Women's greater vulnerability to disasters**

In general, around the world, women are poorer than men. Women are disproportionately employed in unpaid, underpaid and non-formal sectors of economies. Inheritance laws and traditions, marriage arrangements, banking systems and social patterns that reinforce women's dependence on fathers, husbands, and sons, all contribute both to their unfavourable access to resources and their lack of power to change things. The health dangers that result from multiple births can contribute to interrupted work and low productivity. Traditional expectations and home-based responsibilities that limit women's mobility also restrict their opportunities for political involvement, education, and access to information, markets and a myriad of other resources - the lack of which, reinforces the cycle of their vulnerability[16].

Gender-based inequalities interact with social class, race and ethnicity, and age, which put some women and girls especially at high risk. Gender inequalities with respect to enjoyment of human rights, political, and economic status, land ownership, housing conditions, exposure to violence, education and health (and in particular, reproductive and sexual health), make women more vulnerable before, during, and after disasters.

During disasters, there are many casualties among women, especially if women do not receive timely warnings, or other information about hazards and risks, or if their mobility is restricted, or otherwise affected due to cultural and social constraints. Gender-biased attitudes and stereotypes can complicate and extend women's recovery; for example, if women do not seek or do not receive timely care for physical and mental trauma experienced in disasters. Women's relative longevity compared to men's and their reproductive roles can affect mobility and health. Older women (the very old in particular), women with disabilities, pregnant and nursing women, and those with small children are often most at risk, left behind or left out, or the last to leave in cases of emergency departure because they lack knowledge, mobility and resources.

Clearly, high rates of female poverty are an important factor in increasing women's risk in disasters. Also, women's human rights are not comprehensively enjoyed throughout the disaster process. In addition, their economic and social rights are violated in disaster processes when mitigation, relief, and reconstruction policies do not consider the different capacities and needs of men/women, or their differential impact. The right to adequate healthcare is violated when relief efforts do not meet the needs of specific health needs (physical and mental) of women throughout their life cycle, and especially, when trauma has occurred. The right to the security of persons is violated when women and girls are victims of sexual and other forms of violence while in relief camps or temporary housing. Civil and political rights are denied if women cannot act autonomously and participate fully at all decision-making levels in matters regarding mitigation and recovery.

Of course, detailed gender analysis is necessary to discover the exact status of men and women in any given context, including disasters. Such analysis can expose gender-based vulnerabilities, and support the formulation of equitable policies with regard to disaster management and long-term integrated development. This will make mitigation coincide with the ground realities of a disaster situation from the very inception of a disaster intervention.

There are a number of gender analysis frameworks utilized for development planning and practice[17]. Highlighted in the following pages are three frameworks that are especially effective in disaster mitigation efforts.

[16] Excerpted from Anderson M, 'Understanding the disaster-development continuum': in Women and Emergencies, Walker B, ed, *Focus on Gender*, 2/1, Oxfam, 1994

[17] See Candida M, Ines S, Mukhopadhyay M, *A Guide to Gender –Analysis Frameworks*, Oxfam Publications, 1999.

Vulnerabilities and Capacities Analysis of Disaster Situations

The gender-disaggregated Capacities and Vulnerabilities Analysis (CVA) Matrix presented by Anderson and Woodrow (1989)[18] helps to capture most of the subtle but vital gender considerations, which need attention in disaster management and in development plans.

Gender Disaggregated Capacities and Vulnerabilities Analysis Matrix

	Vulnerabilities		Capacities	
	Women	Men	Women	Men
Physical/ Material				
Social/ Organisational				
Motivational/ Attitudinal				

Figure 1

The CVA matrix helps to recognize the gender-based vulnerabilities and capacities that are related to physical/material resources, social/organizational resources, and people's beliefs, motivation, and attitudes.

[18] See Anderson MB, Woodrow PJ *Rising from the Ashes: Development Strategies in Times of Disaster*, Westview press, 1989, pp 9-25 for a full explanation of the Vulnerabilities and Capacities Framework

Given that disasters and emergencies can be extremely complex situations, five other dimensions are added to the above analysis to further reflect this complex 'reality':

- **Disaggregation of communities by gender** to encompass the different needs and interests of women and men.
- **Disaggregation according to other dimensions of social relations** to find out how different groups of people

(distinguished by class, political affiliations, language, age, etc.,) are affected by crisis and interventions
- **Changes over time** (especially before and after crisis intervention) to assess social change and evaluate impact.
- **Interactions between the various categories of analysis** are necessary, because different capacities and vulnerabilities are related to each other. And, changes in one will impact on others.
- **Analysis at different levels and scales of society can** assess the disaster-proneness and the development potential of each level as well as the links between the different levels of societies.

People-oriented Planning Framework

The people-oriented planning framework (POP), an analytical framework prepared for use in refugee situations, emphasizes a number of key factors that require attention in the design and planning of interventions:

Change - In a crisis situation, people's circumstances change rapidly and dramatically. These changes in circumstance may sometimes result in changes to the accepted roles and norms (In some situations traditions may prevail).

Participation of the affected people (men/women/children) in planning mitigation interventions (whether they are temporary or short or longer term) is crucial for the success of the intervention.

Analysis – The POP framework also emphasizes that socio-economic and demographic analysis are crucially important factors for the planning process (irrespective of the type of project intervention).

[19] Ariyabandu MM, 'Framework for incorporating gender issues in a livelihood and disaster context', prepared for the 'Livelihood Options for Disaster Risk Reduction Project' unpublished note, ITDG South Asia, Colombo, 2001

> Components of the POP framework are:
> The **Determinants Analysis** (or Refugee Population Profile and Context Analysis) involving the
> - Population profile of displaced groups and their host community/country (including a gender profile),
> - Social/cultural context that will be influenced by refugees and hosts,
> - Gender division of use and control of resources.
>
> The **Activities Analysis** involves examining who does what, for how long, when, and where, as well as the resources used - in situations where 'normal' life has been disrupted through flight.
>
> The **Use and Control of Resources Analysis** involves discovering how resources are distributed and who has control over their use. Resources provided to refugees in camps are also analysed according to gender and other needs.

The above analysis can capture the changes taking place in refugee situations (especially in relation to gender roles and relations), and provide opportunities of positive change for women.

ITDG South Asia[19] has developed a framework to analyse gender-based division of labour, capacities and vulnerabilities, and specific needs in the disaster cycle. It aims to capture the capacities and vulnerabilities prevalent throughout the disaster cycle, which is closely and intricately linked to the daily livelihood maintenance of the communities living in hazard-prone areas (See matrices in figures 2 - 4).

Framework for gender-livelihoods analysis in disasters

Gender division of labour

Time phases	Economic activities related to livelihood		Non-economic activities related to livelihood (sustenance of family, sustenance of community, collective activities)	
	Women	Men	Women	Men
General/'normal' situation				
In preparation for prevalent disasters in the area (drought, flood, cyclone etc.)				
During disaster/ emergency situations				
Post disaster situation in re-building livelihoods				

Figure 2

Gender-based capacity assessment

Specific skills and capacities displayed during the disaster cycle	Women	Men
Preparedness		
Emergency management		
Re-building livelihoods		

Figure 3

Gender-based vulnerability analysis
(During disasters and emergencies)

	Vulnerabilities	Specific needs
Girl children		
Young women		
Pregnant and lactating women		
Old women		
Old men		
Disabled men		
Disabled women		

Figure 4

Impact of disasters on women and men
Realities for South Asia

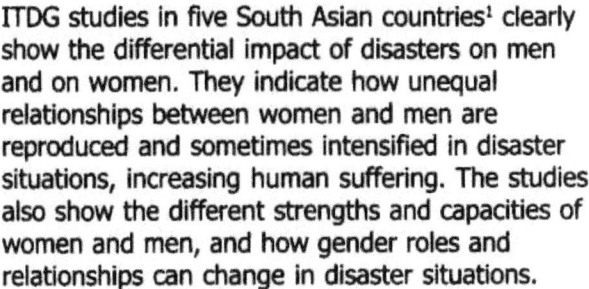

ITDG studies in five South Asian countries[1] clearly show the differential impact of disasters on men and on women. They indicate how unequal relationships between women and men are reproduced and sometimes intensified in disaster situations, increasing human suffering. The studies also show the different strengths and capacities of women and men, and how gender roles and relationships can change in disaster situations.

3.1. Differences in social and cultural impacts

Usually, the impact of disasters is measured in quantifiable ways - such as adding up the numbers of the dead and injured, and estimating the physical damage to housing, land, livestock, agriculture, stores, and infrastructure. But attention is not necessarily paid to how disasters impact on different categories of people: men, women, children, aged people etc.

Disasters affect women and men differently, because of the different roles they occupy; and the different responsibilities given to them in life; and because of the differences in their capacities, needs, and vulnerabilities. The following examples from different South Asian countries convey that most of the gender-related differences in disaster situations arise from the different roles and responsibilities that women and men undertake in their day to day lives.

In situations of disaster, the gendered division of labour becomes critical as gender roles are often seen to be re-enforced and even intensified - due

[1] ITDG South Asia, Research studies on 'Gender issues in Disasters' carried out for the regional project 'Livelihood Options for Disaster Risk Reduction in South Asia', ITDG South Asia, 2001

to the additional work and changes in environment brought on by a disaster.

- The situation in the areas hit by drought[2] in India, shows the reinforcement of the role of women as the caretakers in households. They take on the full responsibility of their nurturing role by waking up at midnight, to trek 5 km or more to a well, and waiting till dawn for a few drops of water. Sometimes, they get nothing. This study notes that these Rajasthani women have no choice but to scrounge for water.

- In Bangladesh "traditional gender specific work such as carrying water, cooking, caring for children and animals became so difficult for women during flood conditions that their lives were at risk ... Often there was no alternative, because there were no men around to help, and even if there were, they did not assist with women's work because of the powerful ideas of gendered division of labour".[3]

However, there are rare occasions when gender roles may change, and even interchange between men and women when responding to a crisis situation.

- During the drought in the village of Andarawewa, Sri Lanka, households began to rely more and more on men to ferry water on pushbikes and tractors from the water bowsers provided by the local government authority.[4]

- During the drought in Tharparkar, Pakistan, when water sources were drying up, men began to share in the responsibility of collecting the larger quantities of water needed for animals.[5]

As discussed earlier, women's reproductive work (related to household maintenance), despite its arduous load, continues to be unacknowledged in

---

[2] Brijnath R, 'Depths of Despair', India Today, May 8, 2000

[3] Nasreen M, 'Coping Mechanisms of Rural Women in Bangladesh During Floods; a Gender Perspective in *Disaster: issues and gender perspectives*, Ahmed N, Khatun H eds, Department of Geography and Environment, University of Dhaka, 2000, p 314

[4] Kottegoda S, 'A study of gender aspects of communities living with drought and landslides in Sri Lanka' carried out for ITDG South Asia, Colombo, 2001, p.28

[5] Waheed A, Sheikh R, 'Gender and Livelihood Options for Disaster Risk Reduction, Tharparkar, Pakistan', Case Study carried out for ITDG South Asia, Journalists Resource Centre, Islamabad, 2000, p 43

societies. In disaster situations, the already heavy workload increases.

- During the drought in Tharparkar, Pakistan, women had to spend more and more time on collecting fuel-wood, water, and in the preparation of food; as water sources and other food and energy sources began to dry up.[6]

- In Rajasthan, India, during drought, women who stay back without migrating end up doing wage work for drought relief programs implemented by the government or NGOs. However, the work is very strenuous, and the villagers say that even if men did not migrate, men would not be ready to take up this kind of work. Thus, more women end up doing drought relief work, which invariably involves the construction of roads and other village infrastructure.[7]

- During floods in Jhang, in the Punjab Province of Pakistan, "when affected people come back to their destroyed houses, male members usually start rehabilitation work on agricultural land, and in caring for livestock. Female members share with them the handling of animals and actively work in the rebuilding of houses, preparing mud and doing construction work, in addition to other household activities."[8]

- In Bangladesh, it is noted that although floods affect all men and women, the physical burden of coping falls heaviest on women, and particularly on women from the poorer categories. During the floods, women's daily household activities were extended to include the extra responsibilities of protecting their households, family members, livestock and other belongings.[9]

[6] Op. cit.

[7] Krishnaswamy PB, Kumar S, Dave M, 'Gender Issues in Livelihood Options for Disaster Risk Reduction' Case study carried out for ITDG South Asia, Disaster Mitigation Institute, Ahamadabad, 2001, p 17

[8] Hameed K, 'Gender issues in livelihoods and flood disaster; case studies on Kamra and Kot Murad Villages, Jhang District, Punjab, Pakistan', carried out for ITDG South Asia, Journalists Resource Centre, Islamabad, 2001, p 31

[9] Nasreen M, 'Coping Mechanisms of Rural Women in Bangladesh During Floods; a Gender Perspective' in *Disaster: issues and gender perspectives*, Ahmed N, Khatun H, eds, Department of Geography and Environment, University of Dhaka, 2000 pp 311-324

- Also during floods, women suffered from lack of food, clothing and shelter. Unemployed men often sat idle or moved elsewhere, leaving their household members behind. It was left to the women to take the responsibility for protecting children and other members of the family, livestock, and belongings.[10]

Family size may change at household levels

- increasing in the wake of an earthquake/floods or decreasing due to migration in drought periods etc. Either way, these changes affect women most - whether in terms of additional mouths to feed and the sick/elderly to care for; or in terms of financial loss, protection and assistance to the household.

- In the Chitwan district, Nepal, during the floods, the extended family sometimes collapsed - leaving the women and elderly without support.[11]

- In some parts of Tharpakar, as drought conditions advanced, men migrated to the barrage areas and to cities. Women were left in the village households to take care of the children, sick and old.[12]

- In the village of Sonari in the drought-stricken Thar in Pakistan, it was observed that only the women, children, and old people remained in the village. Women fetched brackish water from nearby wells and climbed trees to pluck a local fruit that helps to quench the thirst of humans and cattle. The report further noted that without a single man visible in the village, the lives of these hardy women revolve around their donkeys, goats and malnourished children.[13]

---

[10] Op. cit. p 314

[11] Centre for Policy Studies, 'Gender Issues in Livelihood and Disasters - A Case Study of Flood in Nepal' carried out for ITDG Nepal, 2001, p 15

[12] The News, Pakistan 14.5.2000

[13] Op. cit.

- Changes in the household size were reported in Surendranagar, Gujarat, India, following both a drought and an earthquake. Family size was seen to increase more in the wake of the earthquake, than as a result of drought.[14]

If men die, or are injured, or migrate or become involved in relief work, women end up becoming the sole breadwinners for their families. This may result in an increase of female-headed households.

- Bal Kumari Rai's husband died in the 1993 floods, in Nepal. After escorting the family to safer ground, he had returned to the house to save their livestock. He drowned in a swamp. Now the main source of livelihood for Bal Kumari and her five children is wage labour. Although she gets support from her older sons, it is not enough to bring up the three younger children. As a result, she has started selling drinks and working on the "Food for Work" programme, while trying to get hold of some land that she can cultivate, and in which to graze animals.[15]

- Shakhida Begum from Kutubdia Island, Bangladesh, lost her husband to the April 1991 cyclone. Her house was washed away, and she lost her livelihood (of working in the family saltpan), after her stepson encroached onto the saltpans. In order to support her three children, she has taken to begging with her two sons. Being a saltpan area, the alms they get are usually in the form of salt, which they then sell in the market, or to moneylenders to get cash to support their needs.[16]

As noted earlier, during or after a disaster, as job opportunities dry up, men have the option of migrating to find work. Women are less able to migrate because of their care-giving

[14] Enarson E, 'We want work: Rural Women in the Gujarat drought and earthquake', Report based on a Quick Response Grant from The Natural Hazards Research and Applications Centre and the National Science Foundation, www.colorado.edu/hazards/qr/qr135/qr135.html

[15] Centre for Policy Studies, 'Gender Issues in Livelihood and Disasters - A Case Study of Flood in Nepal' carried out for ITDG Nepal, 2001, pp 24-27

[16] Kafi SA, *Disaster and Destitute Women, twelve case studies*, Bangladesh Development Partnership Centre, Dhaka, 1992, pp 42

[17] Krishnaswamy PB, Kumar S, Dave M, 'Gender Issues in Livelihood Options for Disaster Risk Reduction' Case study carried out for ITDG South Asia, Disaster Mitigation Institute, Ahamadabad, 2001 p 17

[18] Kafi SA, *Disaster and Destitute Women, twelve case studies*, Bangladesh Development Partnership Centre, Dhaka, 1992, p 23

[19] Op. cit.

responsibilities, which may keep them housebound. This makes them more prone to health hazards as well as the other vulnerabilities arising from the harsh conditions of disaster.

- A study in Rajasthan, India, found that during drought, a large number of men migrate in search of work, while most of the women stay behind in the villages. This puts an added burden on women, especially when the availability of adequate nutrition and clean drinking water dwindles. This affects their own health and the health of family members.[17]

- Since women are unable to find work in the area, Khodeja from Tablar Char, Kutubdia, Bangladesh, has had to send her children begging after she lost her husband and house to a cyclone.[18]

At the same time, social practices and cultural values of the region and community can also influence how women and men are impacted by a disaster, and how they react to its effects. Often, the prohibitions of some cultural/social practices restrict women more than men. These limits are imposed through male authority, as well as through personal self-limitation that is practised by women.

For instance, the tradition of Purdah (veil) limits women's mobility in many parts of South Asia and restricts women's access to information. As a result, sometimes, women do not receive disaster warnings; and may be less able to take action even when they do. Studies in Bangladesh show examples of how women did not respond to flood or cyclone warnings because of these problems of seclusion and mobility.[19]

- Khodeja Begum narrated that in the night of the cyclone, she did not go to a cyclone shelter, as there were none close to her house.

Rather, she opted to take refuge, along with her children, on the high, homestead ground of a neighbour.[20]

- Women, especially young women, did not feel comfortable, and were unwilling to go to the cyclone shelters because of certain cultural values. They expressed concern that exposing themselves to various men would have negative impacts on their social status within the family, as well as within their kinship group.[21]

- The study in Bangladesh conveyed how girls and women were exposed to dangers other than those arising from the cyclone. There was a high risk involved in going from place to place searching for shelter, as there were some 'anti-social elements' hunting for young girls and women in order to take advantage of the situation.[22]

- In Bangladesh during floods, women depend largely on their male counterparts to decide when and whether to leave home for a safer place. It is usually the man's decision, which finally makes a family move.[23]

In some situations, women and girls have less chance of surviving a disaster because of attitudes and practices that discriminate against them; and give them a lower social status than males.

- Wide gaps in social, demographic and economic indicators point to the lower status of women in Bangladesh, when compared to men. Within the family unit too, women and girls are subject to discrimination. It is probable that attitudes such as male preference lead to such practices as the inclination to save the male child, or male family members first, during a disaster.

[20] Op. cit. p 28

[21] Rahman MA, Rowshan D, 'Gender and Natural Disasters in Bangladesh: Strategies to Reduce Vulnerability of Women', Report on the South Asia Consultation on Natural Disaster Management, Rajapakse S, Bhagbanprakash Eds, Ministry of Youth Affairs and Sports, Government of India, Commonwealth Youth Programme: Asia Centre, Bhuwaneswar, India, Nov 29- Dec 3, 2001, p 4

[22] Kafi SA, *Disaster and Destitute Women*, twelve case studies, Bangladesh Development Partnership Centre, Dhaka, 1992, p 14

[23] Saleheen M, Khondaker M, Huda S, 'The Vulnerability of Women in Disaster Prone Areas in Bangladesh' in *Disaster: issues and gender perspectives*, Ahmed N, Khatun H eds, Department of Geography and Environment, University of Dhaka, 2000, p 100

Female family members may be deprived of relief due to male family members controlling access to health care, food and other emergency resources.[24]

- It was observed that during the relief and rehabilitation phase following the cyclone and other disasters in Bangladesh, families headed by women had little or no chance to get at relief supplies. Whereas, families with more male members had greater access to the scanty relief sources.[25]

- "Abul Kalam had five daughters and one son. He was a poor sharecropper. He was holding his children together and fighting against the wind - fearful of the rising water. In his struggle to survive, Abul Kalam released his daughters, one after the other, so that his son could survive."[26]

Often women and girls in South Asia are brought up to sacrifice their own comfort and prospects in favour of their male family members. Men in the families are prioritized, while girls and women may eat last, and least, and become undernourished, thereby reducing their chances of survival.

- The study on floods in Pakistan showed that "In cases of regular flooding in the region (Jhang District, Punjab), food items become scarce mainly due to the devastation of agricultural land. Women and girls received less nutrition, as the cultural priority is to first serve food to men and boys. This results in poor physical growth, and greater risk of poor health in women."[27]

- After the floods in Chitwan, Nepal, girls did not have enough time to go to school because they were forced to walk further and further to collect fodder and firewood for their

---

[24] Rahman MA, Rowshan D, 'Gender and Natural Disasters in Bangladesh: Strategies to Reduce Vulnerability of Women', Report on the South Asia Consultation on Natural Disaster Management, Rajapakse S, Bhagbanprakash eds, Ministry of Youth Affairs and Sports, Government of India, Commonwealth Youth Programme: Asia Centre, Bhuwaneswar, India, Nov 29- Dec 3, 2001, p 5

[25] Op. cit, p 4

[26] Enarson E, Gender and Natural Disasters, Infocus programme on Crisis Response and Reconstruction, Working paper 1, Recovery and Reconstruction Department, ILO, Geneva, 2000, p 4

[27] Hameed K, 'Gender issues in livelihoods and flood disaster; case studies on Kamra and Kot Murad Villages, Jhang District, Punjab, Pakistan', carried out for ITDG South Asia, Journalistic Recource Centre, Islamabad, 2001 p 19

families. Some girls were forced to discontinue their education. (While some of the fathers spent their time playing cards and drinking alcohol, and not earning for their families).[28]

- In Gujarat India, after the earthquake, it was found that girls were burdened by the increased responsibility of having to care for their siblings, while their mothers searched for wage work. They also had to work harder at domestic tasks in living conditions made much more difficult by the quake.[29]

Social perceptions about widows and women living without 'male protection' as well as cultural taboos pertaining to women can result in social marginalization

- In Rajasthan, India, "Drought has created loss of work in the villages, and men have migrated to the cities creating a large number of women-headed households. Women are living alone in the villages ... In a few villages, incidences have been reported of women being harassed and exploited by upper caste people. This is forcing these women to remain confined to their homes and villages. They do not feel comfortable to go out to distant places of work and are scared to collect water late at night." [30]

- Sabmeraj of Akyab, who has come from Maynamar as a refugee, at the Cox's Bazaar, Bangladesh refugee camp narrated that – "As we do not have a man among the family members it took time (18 days) to get our names on the refugee list. Those who had men in their family, they were very fortunate to have their names on the official refugee list at the first chance." [31]

- In Tharparkar, Pakistan, boys are left behind with neighbours to continue their education,

[28] Centre for Policy Studies, 'Gender Issues in Livelihood and Disasters - A Case Study of Flood in Nepal' carried out for ITDG Nepal, 2001, p 15

[29] Enarson E, 'We want work: Rural Women in the Gujarat Drought and Earthquake', Report based on a Quick Response Grant from The Natural Hazards Research and Applications Centre and the National Science Foundation, 2001, www.colorado.edu/hazards/qr/qr135/qr135.html, p 16

[30] Kumar S, 'Livelihood and drought, a gender differential, Rajasthan', Centre for Disaster Management, Rajasthan State Institute of Public Administration, Jaipur, India, 2001, p 13

[31] Kafi SA, *Disaster and Destitute Women*, twelve case studies, Bangladesh Development Partnership Centre, Dhaka, 1992, p 20

when entire families migrate to barrage areas, whereas it may be difficult to leave behind girls due to safety concerns.[32]

At the same time, it must be noted, that there are occasions when certain cultural practices/taboos become less prominent during disasters.

- During floods in Bangladesh, "The strong gender division of labour forced poorer women to do certain tasks for their households, which led them to compromise on *purda*. Women in poorer categories often acted as men did - in making platforms, cutting bamboo, making bamboo bridges, protecting crops and livestock and engaging in income generating activities. Men, on the other hand, remained more restricted to their gender-assigned activities."[33]

3.2 Differences in economic impacts

At a broad level, disasters affect women and men alike. Very often, a disaster may mean that people have to start their lives from scratch. Assets are destroyed; crops fail, and household possessions are lost, while livestock is usually the first casualty.

In some instances, disasters turn into a downward spiral. This is when a community that has already been affected by one disaster (where assets have been destroyed, or sold off in order to recover), is struck by another disaster; compelling the affected community into a worse situation – until the next disaster; depleting their meagre resources over and over again. The situation in Gujarat, India, amply demonstrated this in January 2001, when an earthquake hit communities already affected by drought.

---

[32] Waheed A, Sheikh R, 'Gender and Livelihood Options for Disaster Risk Reduction, Tharparkar, Pakistan', Case Study carried out for ITDG South Asia, Journalists Resource Centre, Islamabad, 2000, p 12

[33] Nasreen M, 'Coping Mechanisms of Rural Women in Bangladesh During Floods; a Gender Perspective' in *Disaster: issues and gender perspectives*, Ahmed N, Khatun H eds, Department of Geography and Environment, University of Dhaka, 2000 p 316

Nor is it easy or sometimes possible to recover from a disaster. Despite the common perception of disasters as being sudden and short-term; to be tackled with a quick fix; its impact on people and their livelihoods may be extensive. Natural resources may be depleted or exhausted to such an extent that people's livelihoods may be permanently damaged.

For instance, floods may wash away fertile soil, leaving the land fallow for years. Landslides may permanently destroy lands and habitats. Droughts may go on for years, or they may be compounded by another disaster such as an earthquake. Consequently, there can be an urgent need for alternative employment and income generation opportunities.

As employment patterns change and livelihoods are lost, there is an overall sense of economic insecurity. Women are the worst hit as they already live on the fringes of poverty - even before the disaster. In relation to men, women's role in the subsistence economy is particularly significant as economic actors whose time, efforts, and incomes are used to sustain life for the others in their families.

Studies on South Asia show that, contrary to the popular perception that women were not involved in productive work, women actually do work in their own and other people's lands; and are responsible for a variety of tasks that include threshing, cleaning, drying, storing and growing of vegetables and winter crops, feeding the cows and poultry, re-plastering huts with mud, stitching and mending quilts and mats, and a host of other jobs. Floods and drought deprive these women of their jobs and incomes. The loss of this revenue means lesser quality of life for their families, and impact on the overall recovery of the community after a disaster.

Greater lack of income opportunities for women

Even in 'normal' situations, women are already disadvantaged because of the lack of opportunities. Although wage opportunities decrease for both men and women due to loss of land after a disaster, men have the option of migrating and better chances of finding work in restoration. Women stay behind with their families.

- Women affected by floods in Chitwan, Nepal spoke of their limited options for employment and income generation: " ... the main common problem of women and young girls is the lack of employment opportunity for earning a livelihood. Women who had lost all their meagre belongings and their life savings, have not been able to compensate their losses even after seven years."[34]

- A study on Gujarat, India, records that in many drought-impacted households across Gujarat, women are in need of paid work to support families (since the men have left the villages in search of wage work). However, conflicts with childcare and other domestic work, gender restrictions on tasks and occupations, low wages, and hazardous working conditions limit their ability to convert hard work into income.[35]

[34] Centre for Policy Studies, 'Gender Issues in Livelihood and Disasters - A Case Study of Flood in Nepal' carried out for ITDG Nepal, 2001 p 14

[35] Enarson E, 'We want work: Rural Women in the Gujarat Drought and Earthquake', Report based on a Quick Response Grant from The Natural Hazards Research and Applications Centre and the National Science Foundation, 2001, p 9 www.colorado.edu/hazards/qr/qr135/qr135.html

Loss of women's assets

Households facing economic crisis become more dependent on women's incomes and resources in the aftermath of natural disasters. Women may be forced to sell their own personal assets such as jewellery in order to secure food for the family.

- Studies in Pakistan and Sri Lanka[36] record that women pawn or sell their jewellery (sometimes, this may be the only personal property owned by women in the sub-continent).

- After a disaster, women sometimes take loans to recuperate; and end up becoming long-term labourers, working off their debts to moneylenders.[37]

- Home-based piece-rate workers and self-employed women lose their space of work, tools and other resources as well as their clientele.[38]

- A study in Rajasthan, India, found that, "until drought relief work starts, the family often sells or pawns ornaments and utensils, and women often provide financial support from their savings. Women's assets are the first to be disposed of." [39]

- In Bangladesh, women's assets (poultry, kitchen utensils, goats, milch cows) were used to meet the immediate needs of the households during floods, whereas men's assets (animals for cultivation) were kept for the future. Women's assets are more easily converted to cash, and are therefore, the first to be disposed of, so as to meet the immediate needs of family survival.[40]

[36] Hameed K, 'Gender issues in livelihoods and flood disaster; case studies on Kamra and Kot Murad Villages, Jhang District, Punjab, Pakistan', carried out for ITDG South Asia, Journalists Resource Centre, Islamabad, 2001 and Kottegoda S, A study of gender aspects of communities living with drought and landslides in Sri Lanka' carried out for ITDG South Asia, Colombo, 2001

[37] Op. cit.

[38] Op. cit.

[39] Krishnaswamy PB, Kumar S, Dave M, 'Gender Issues in Livelihood Options for Disaster Risk Reduction' case study carried out for ITDG South Asia, Disaster Mitigation Institute, Ahamadabad, 2001, p 17

[40] Nasreen M, 'Coping Mechanisms of Rural Women in Bangladesh During Floods; a Gender Perspective' in *Disaster: issues and gender perspectives*, Ahmed N, Khatun H eds, Department of Geography and Environment, University of Dhaka, 2000, p 316

[41] Bhutta AH, 'The Response of Riverine Communities to Disasters; A case study of a Pakistan Village Mamola, with Special Reference to Changes in Livelihood Patterns and Community Based Rehabilitation'. Unpublished report, ITDG South Asia, 1999

[42] Kottegoda S, 'A study of gender aspects of communities living with drought and landslides in Sri Lanka' carried out for ITDG South Asia, Colombo, 2001

[43] Op. cit.

- Jewellery, in South Asian societies, is a very significant asset for women. It is part of the dowry that provides security to women. Jewellery is seen as insurance for old age; and in the eventuality of abandonment, widowhood, and emergencies; and women cling to these meagre assets throughout their lives. After the crisis, in the riverine communities of Pakistan, as women parted with their jewellery, they talked of how this made them lose the only personal security they had; increasing their vulnerability, as they often did not have a method of replacing these assets.[41]

Women's increasing economic dependency on men

In some cases, women's economic dependence on men may increase further.

- In Anamaduwa, in the Puttalam District of Sri Lanka, during the drought, the number of men cultivating crops as well as the number of crops being cultivated decreased. No women were recorded as being engaged in the cultivation of crops during the drought. And the only produce that women continued to rely on for an income was fruit.[42]

- When a drought hit Hambantota and neighbouring districts in the south of Sri Lanka, paddy cultivation suffered, and women lost the opportunity to work as wage labourers. However, men (who also lost their incomes from paddy cultivation) had the option of migrating to the towns and cities for work.[43]

Increased suffering amongst vulnerable groups

Poor and disadvantaged groups are more vulnerable during disasters and in the aftermath; and suffer more acutely due to their disadvantaged situations.

- Work opportunities for women in many areas in Bangladesh are virtually non-existent. Therefore, when they are hit by disasters and displaced, many families are forced to take to the begging bowl for survival. The situation is worse for women who do not have a male relative in the family (husband, a brother or a grown-up son). There are numerous accounts of the children of female-headed families affected by disasters. These children are never sent to school, but instead, sent out to beg for family survival.[44]

- Furthermore, due to the cultural restrictions on women's mobility, they are unable to move to unknown destinations on their own, in search of employment. Being illiterate, the skills possessed by these women are limited and basic.[45]

- "It has been found that during floods in Orissa a large number of women and children drown. When they survive disasters, in destitute conditions, younger women are forced into prostitution and older women turn to begging and charity." [46]

[44] Kafi SA, *Disaster and Destitute Women*, twelve case studies, Bangladesh Development Partnership Centre, Dhaka, 1992, pp 42-43

[45] Ariyabandu MM, 'Impact of Hazards on Women and Children; Situation in South Asia' Conference proceedings, reaching Women and Children in Disasters, Miami, Florida, 2000. www.anglia.ac.uk/geographygdn

[46] Krishnaswamy PB, Kumar S, Dave M, 'Gender Issues in Livelihood Options for Disaster Risk Reduction' Case study carried out for ITDG South Asia, Disaster Mitigation Institute, Ahamedabad, 2001, p 13

3.3 Differences in psychological impacts

The psychological effects of disasters on men and women are usually not recorded or taken into account in mitigation efforts. Case studies tend to convey that women are more psychologically affected by disasters than men. This is due to the fact that usually women are not only forced to take on the roles of frontline responders and survivors, but also caregivers during and after a disaster. This is owing to the simple reason that women generally feed other people and help to keep them healthy and strong – thereby reducing their family members' vulnerability to disaster.[47] Thus women's anxiety also stems from fear/perception of risk to family/children.

Here, it is possible to observe how there are fewer expectations of men or even different demands made on their time and labour as opposed to women.

- A relief worker in Bangladesh comments, "Because coping with poverty is harder for women in general, the aftermath of the cyclone hit them the hardest. Their men may have lost the fishing equipment necessary to earn a living, their children may have died; and their home and belongings washed away, but at the end of each day, it was the wife/mother who had to cook for whoever who had survived in her family ... They were collecting the bits of wood and bamboo to rebuild the houses. As is customary, they dealt with the sick children and lack of food".[48]

[47] Bhatt M, Gender and Disaster: Perspectives on Women as Victims of Disaster,' Discussion Paper, Disaster Mitigation Institute, Ahamadabad, India, 1995

[48] Bari S, 'Women in the Aftermath' in *From Crisis to Development; Coping with Disasters in Bangladesh*, Hossain H, Cole PD, Abel FH eds, Dhaka University Press Ltd., 1992, pp 55-58

The breakdown of household and social structures during disasters may make women more vulnerable to stress and anxiety.

- After the 1993 floods in Chitwan, Nepal,[49] women were seen to suffer from post disaster reactions such as nightmares and horror of future floods, for many years after the event.

Women usually do not get time to mourn the loss of family members; and thus, suffer from post-traumatic stress disorder in comparatively larger numbers than men. Women may also be victims of post-partum strain when disasters strike immediately after childbirth.

- "At the time of the flood, Shanthi was at her post-partum stage of 10 days. She required good food and rest, but had to remain in the camp along with her neighbours. She had to clean, take care of young children, and work as a manual labourer along with her fellow women. Though she received some assistance in the form of old clothes, cereals, utensils, etc., from relief agencies, she was not given the special care required for a post-partum woman." [50]

Emotional stress for women heightens as the resources needed to feed and support a family are depleted through disaster; the feeling of not being able to meet their responsibility adds to their helplessness and depression. At the same time, biological experiences such as childbirth have certain repercussions and require special care that are ignored in disaster interventions.

- "Thuli had delivered 10 days before the flood. At a time when she needed complete rest, good care and nutritious food for the physical upkeep of both mother and child, she had to

[49] Centre for Policy Studies, 'Gender Issues in Livelihood and Disasters - A Case Study of Flood in Nepal' carried out for ITDG Nepal, 2001

[50] Op. cit. p 21

[51] Op. cit. p 22

[52] Op. cit. p 17

[53] Op. cit.

[54] ITDG South Asia, 'Notes from the Focus Group discussions with a community affected by rockfall in Nawalapitiya, Sri Lanka', 2001

stay in a relief camp. After four days in the camp, the baby caught fever. She was not able to satisfy the child by breast-feeding, and the family had no other means to feed the child. Thuli was also very weak and often had nausea. She also had to take care of their 6 other children. Her husband, highly depressed over the loss of property and the pathetic condition of his wife, was not in a mood to search for work." [51]

- "Shanti Maya stayed in a camp for about a month along with other flood-affected families in a congested environment. During her stay there, she was under severe mental stress and strain, due to the feeling of uncertainty and loss of confidence. She was worried about the loss of property, livestock, crops and land. The environment in the camp was painful not only because everyone was distressed, but also because of physical ailments. There were cries from every corner all through the day and night. Shanti Maya felt suffocated, but had no options.' [52]

As noted before, while men may migrate in the aftermath of a disaster, women become the frontline fighter/strategist who stay at home; thereby, intensifying their sense of entrapment and lack of options. The cyclic nature of some disasters is a source of constant emotional pressure in general.

- In the Chitwan district, Nepal, during the rainy season each year, villages are haunted by the fear of the possible breakage of embankments along the river Rapti.[53]

- More women recorded mental anxiety at the prospect of harm to their families through landslides in Nawalapitiya, in Sri Lanka.[54]

Increase in Violence against Women

Studies have also shown that violence against women increases during and in the aftermath of disasters. The incidence of wife battering, rape, sexual harassment and sexual abuse becomes more frequent as the families are stripped of their age-old traditions and mechanisms of protection and support, during disasters.

Particularly in camp situations, there are increased levels of gambling and alcohol use by men as they try to channel their fear and frustration into these activities, and thereby, aggravate the situation.

- Women affected by floods in Chitwan, Nepal, were very concerned about the alcoholism of their husbands. In the absence of employment, many men were using up their time by playing cards and drinking liquor. In trying to advise husbands to refrain from such activities for the sake of family welfare, the women were abused by their men with assertions that they had no right to control their spending of paternal property.[55]

- The study on drought in Rajasthan, in India, showed that tube well water was sometimes issued only in the middle of the night. For some women, the walk to the pump was up to 4 miles. With no other access to water, women had to brave this journey at night, at grave risk.[56]

- There were also incidents in some Rajasthan villages, of women being harassed and exploited by upper caste groups and contractors when they went to work.[57]

[55] Centre for Policy Studies, 'Gender Issues in Livelihood and Disasters - A Case Study of Flood in Nepal' carried out for ITDG Nepal, 2001, p 59

[56] Kumar S, 'Livelihood and drought, a gender differential, Rajasthan', Centre for Disaster Management, Rajasthan State Institute of Public Administration, Jaipur, India, 2001, p 11

[57] Op. cit. p 13

[58] Kafi SA, *Disaster and Destitute Women, twelve case studies*, Bangladesh Development Partnership Centre, Dhaka, 1992, p 14

- After the cyclone in Bangladesh, 'as soon as the water receded, the hooligans and the looters were out again, engaged in looting as well as hunting for young women and girls. They did not even spare the dead bodies.' [58]

Recognizing community capacities

Major calamities where thousands die make news and go into historical records. Assistance is made visible through the sums donated, the aid workers and volunteers who are mobilized, the equipment assembled, the dry rations distributed, etc. However, many smaller disasters that occur frequently are hardly reported and scarcely given coverage: how they are managed locally; as well as the resilience and capacities shown by the affected communities in facing disasters.

[1] Cuny, FC, *Disasters and Development*, INTERTECT Press, Texas, USA, 1994

In fact, whatever the scale of the disaster, a major share of recovery and rehabilitation is done by the communities themselves, using available resources, knowledge, and skills. Even the larger international disaster responses do not actually amount to more than 30-40% of the total expenditure of disaster recovery, because usually the balance comes from the affected community.[1]

Disasters are unexpected, and give rise to complex and stressful situations, where societies have to deal with the deaths of large numbers of people, as well as numerous cases of injury, trauma, loss of assets, and displacement. The nature of disasters usually leads to a focus on the vulnerabilities of communities, even though it is equally important to recognize the resourcefulness and capacities of those affected.

People display the capacity to cope through various mechanisms. They include family, kinship, social relationships and local support groups. Some coping mechanisms at community level may have evolved over centuries and are often specific to a particular location and culture. Though not always recognized, individual as well as collective coping

[2] Ariyabandu MM, *Defeating Disasters, Ideas for Action*, ITDG Sri Lanka for Duryog Nivaran, Colombo 1999.

mechanisms assist people in their struggle for survival in the immediate aftermath of a disaster as well as in the long term.

Consequently, the effectiveness of any outside assistance depends on identifying and building on people's social and cultural systems, including their own coping mechanisms.

4.1 Gender-based differences in coping with disasters

As noted earlier, coping mechanisms take many forms; and men and women employ different coping mechanisms. There are strong gender-based differences in the capacity of a coping mechanism to assist men/women deal with a situation, as well as in the in the way that a coping mechanism is employed.

Due to the gender division in labour, men and women possess specialized skills and strengths to cope with crisis. Yet, women's roles in mitigating and preparing for disasters, in rescue operations, as sustainers and re-builders are largely unrecognized, even though their skills and contributions at the household and community levels are crucial.

Thus, it is vital to realize that, contrary to popular perceptions; women are not helpless victims, but display great capacity in extreme situations. They possess skills, resilience, and appropriate coping knowledge.[2]

In fact, women prepare for emergencies; they save lives during crises; and build livelihoods in post-emergency situations. Women therefore constitute an asset; they are a resource. Yet, they are a resource that is unseen and under-used.

4.1.1. Gender-based differences in community preparedness for disaster

Historically, women have used their indigenous knowledge for disaster preparedness; and developed their own coping mechanisms and strategies. They prepare households; try to protect assets and possessions in the event of a disaster; they are, in most cases, a support to others in similar situations, and share information and warnings through their informal networks.

- In the Nawalapitiya township, Sri Lanka, where communities live with the threat of landslides and rock falls, it was observed that women were more likely to witness early signs of landslides or anticipate rock falls, since they stayed/worked around their houses while the men went out to work. They, along with men, formed neighbourhood vigilant groups, during the heavy rains to watch out for potential rock falls.[3]

- In Pakistan, in the Jhang area in Punjab, when preparing for floods, male members of the family take livestock to protective embankments or distant places, and regularly arrange for their fodder. Women make preparations to take care of children, house luggage, valuables, and cooking utensils. It is they who make provision for food, to support the family during the floods, and preserve seed material for the next cultivation season.[4]

- In Rajasthan and Gujarat, (India), in frequently drought-prone areas, women are active in watershed management, and in preserving the water that is collected for domestic needs.[5]

[3] ITDG South Asia, 'Notes from the Focus Group discussions with the selected community groups in Nawalapitiya, Sri Lanka', 2001

[4] Hameed K, 'Gender issues in livelihoods and flood disaster; case studies on Kamra and Kot Murad Villages, Jhang District, Punjab, Pakistan', carried out for ITDG South Asia, Journalists Resource Centre, Islamabad, 2001

[5] Kumar S, 'Livelihood and drought, a gender differential, Rajasthan', Centre for Disaster Management, Rajasthan State Institute of Public Administration, Jaipur, India, 2001

[6] Kottegoda S, 'A study of gender aspects of communities living with drought and landslides in Sri Lanka' carried out for ITDG South Asia, Colombo, 2001

[7] Kumar-Range S, Enarson E, E mail Discussion "Environmental Management and the mitigation of Natural Disasters: a Gender Perspective", United Nations Division for Advancement of Women (DAW), International Strategy for Disaster Reduction(ISDR), November 2001

[8] Op. cit.

- In Sooriyawewa, in the Hambantota District of Sri Lanka, where drought is a frequent hazard, accessing drinking water becomes a major problem for more than 6 months of a year. Usually, women use money from savings schemes to purchase large barrels to store the drinking water that is distributed by the government.[6]

Women's immense knowledge of their living environment is not only used for conservation and thereby to mitigate disasters; but it is also transferred to the next generation. In many parts of the world, girls learn from the older women in their community how to identify plant and animal species and their specific ecological, dietary and health characteristics.

- In a participatory study undertaken in Dehra Dun (India), women identified no less than 145 species of forest plants, (which they manage, and thus have knowledge of), while the forest officer of the area could only identify 20 species.[7]

- Other studies show that women usually prefer multi-species forests which sustain the water cycle, and which offer them diverse products. On the other hand, men are more interested in monoculture forests (which are prone to water logging, and hence, lack soil protection) for profit.[8]

4.1.2 Coping as disaster strikes

Both women and men, and the community as a whole, provide emergency assistance to one another during a disaster - such as calling out warnings to neighbours, and taking children and the sick/elderly to higher places during floods. People also try to save their assets such as jewellery and livestock. Sometimes, this may even cost them their lives.

- *"She was a good woman,"* reported a husband whose wife stayed behind, and was killed in a cyclone.[9]

- Research proves that women are the first to provide nursing-care to the most affected family members, before any official relief work begins. Along with providing this kind of care, it is also beholden on the women to find food, water, and fuel to prepare the next meal for their families. [10]

- In Gujarat, India, after the 2001 earthquake - 'women must still provide food and water for the household and tend to animals who are desperately in need of water and fodder; they cook and clean in their huts, tents or shelters under plastic sheeting, supervise children at play in the rubble or at study in makeshift schools, and care for the ill and the injured out of doors, without ready water or food.[11]

- In Jolpur, Bangladesh, women were responsible for collecting and storing water during the floods. 'Almost all tube wells went under water and poor women had to take considerable risks to procure drinking water from great distances. They had to walk through chest-high water or swim to collect fresh and clean water'. The study also noted that women use various techniques to take

[9] Bari S, 'Women in the Aftermath' in From *Crisis to Development; Coping with Disasters in Bangladesh*, Hossain H, Cole PD, Abel FH eds, Dhaka University Press Ltd., 1992, p 55

[10] ITDG South Asia, Research studies on 'Gender issues in Disasters' carried out for the regional project 'Livelihood Options for Disaster Risk Reduction in South Asia', ITDG South Asia, 2001

[11] Enarson E, 'We want work: Rural Women in the Gujarat drought and earthquake', Report based on a Quick Response Grant from The Natural Hazards Research and Applications Centre and the National Science Foundation, 2001, p 16
www.colorado.edu/hazards/qr/qr135/qr135.html

water out of the tube wells so as not to mix it with flood water, and to purify the water, in the absence of fuel wood.[12]

Also, in procuring food for the family during floods, women mange by gathering edible wild plants, and rotten or discarded vegetables. Frequently, the collection of food from common land was women's work. When men failed to earn any money, women sought other alternatives. They experimented on new plants out of necessity, and quarrelled in the competition for food from common lands. It is also observed that the social networks of women provide emergency survival support during floods. For instance, food items, fuel, bedding, and material for makeshift household purposes are borrowed on loan or given as charity. Borrowing food or small things is not identified as a man's responsibility, as asking for help from others is usually seen as beneath his dignity.[13]

- "Men have time to sit and think – but women have to start their activities immediately. Immediately after the earthquake, she has to prepare food, she has to gather water... Women do not have this time ... Women have to do many activities – men have to do only one or two activities ... She starts from early morning and goes till 8 or 9 at night. Women's work continues...".[14]

- During severe degrees of drought in Rajasthan, women were active in the revival of traditional water sources, securing drinking water for cattle, and taking part in crisis management meetings (AKKAL meeting), and in agitation initiatives.[15]

---

[12] Nasreen M, 'Coping Mechanisms of Rural Women in Bangladesh During Floods; a Gender Perspective' in *Disaster: issues and gender perspectives*, Ahmed N, Khatun H, eds, Department of geography and Environment, University of Dhaka, 2000, p 316

[13] Op. cit. p 322

[14] Enarson E, 'We want work: Rural Women in the Gujarat drought and earthquake', Report based on a Quick Response Grant from The Natural Hazards Research and Applications Centre and the National Science Foundation, 2001, p 16 ww.colorado.edu/hazards/qr/qr135/qr135.html

[15] Kumar S, 'Livelihood and drought, a gender differential, Rajasthan', Centre for Disaster Management, Rajasthan State Institute of Public Administration, Jaipur, India, 2001

4.1.3 Re-building after a disaster

In the immediate aftermath of a disaster, often there are many parties to assist in re-building. In most instances however, external support is limited to taking care of the immediate requirements, and is offered only in the case of large-scale, visible, and publicized disasters. A major part of the re-building of homes and livelihoods after a disaster is done by the communities themselves. This may include a variety of activities, such as land renovation, productivity-raising efforts, sale of land/farms, borrowing and pawning for consumption and investment, agricultural farming, search for wage labour etc.

The responsibility for some of these activities is shared between men and women within a family, yet, gender differentiation is commonly observed in some activities.

Studies convey[16] that while in certain instances, cultural barriers may prevent women from taking an active role in reconstruction, on the whole, women contribute to such activities as relief distribution, clearing-up after disasters, preparation of land, wage labour in reconstruction etc. Sometimes, even despite such cultural barriers as sex segregation, women are seen in queues for food and in reconstruction work.

- Women earned money through relief work (such as soil digging, rubble clearing,) in Gujarat, in regard to both the crisis of drought and earthquake.[17]

- Barely three weeks since the earthquake in Gujarat, India, where an estimated 30,000 were killed and thousands of villages were destroyed, women of Dhaneti village resumed their embroidery work in a makeshift community handicrafts centre built by the Shrujan Trust. According to the women

[16] ITDG South Asia, Research studies on 'Gender issues in Disasters' carried out for the regional project 'Livelihood Options for Disaster Risk Reduction in South Asia', ITDG South Asia, 2001

[17] Enarson E, 'We want work: Rural Women in the Gujarat drought and earthquake', Report based on a Quick Response Grant from The Natural Hazards Research and Applications Centre and the National Science Foundation, 2001, www.colorado.edu/hazards/qr/qr135/qr135.html

artisans, returning to work so soon after the disaster not only gave them some income (Rs 50/- a day) to support their families, but it also helped them to deal with the emotional trauma arising from the disaster.[18]

The work done by grassroots women's organizations contradicts the impression of women as mere victims in disaster situations - especially as this is how women are generally portrayed in the media during disasters.

- A recent initiative in Surendranagar (India) to train village women to manage local water supplies enabled the fast restoration of water after an earthquake. The women concerned mustered 'sufficient courage' to climb on top of a large overhead storage tank to repair a pipe damaged by the quake. The women's action can be contrasted with that of the local mason in the village, who simply fled, as there were frequent tremors in the area.[19]

- In a study undertaken in Rajasthan, India, it was found that " .... drought relief work is mainly done by women, and invariably involves the construction of roads and other infrastructure"[20]

- In Jhang, Punjab, Pakistan, in the aftermath of floods - "When affected people come back to their destroyed houses, male members usually start rehabilitation work on agricultural land and in caring for livestock. Female members share responsibilities with them in the handling of animals and in the rebuilding of houses which involves preparing mud and doing construction work." [21]

- In Punjab, Pakistan, women repair houses as a norm. In disaster situations, they are actively involved in reconstruction efforts and in restoring houses.[22]

---

[18] Reuters Foundation, AlertNet, Relief Resources http://www.alerttnet.org/thefacts/reliefresources/226096.htm

[19] Enarson E, 'We want work: Rural Women in the Gujarat drought and earthquake', Report based on a Quick Response Grant from The Natural Hazards Research and Applications Centre and the National Science Foundation, 2001, www.colorado.edu/hazards/qr/qr135/qr135.html

[20] Krishnaswamy PB, Kumar S, Dave M, 'Gender Issues in Livelihood Options for Disaster Risk Reduction' Case study carried out for ITDG South Asia, Disaster Mitigation Institute, Ahamedabad, 2001 p 17

[21] Hameed K, 'Gender issues in livelihoods and flood disaster; case studies on Kamra and Kot Murad Villages, Jhang District, Punjab, Pakistan', carried out for ITDG South Asia, Journalists Resource Centre, Islamabad, 2001, p 31

[22] Op. cit.

Current practice
The absence of gender-sensitivity in disaster management

The previous chapters highlighted the gender-based differences in disaster impact, the vulnerabilities, and the capacities and needs of men and women which emanate from deep-rooted social, cultural and political relationships. The Dominant Approach to disaster management prevalent in South Asia ignores these differences in planning for the various stages of the disaster cycle, resulting in insensitive and ineffective interventions. This may lead to the perpetuation of the disaster spiral, the further marginalization of women and other vulnerable groups, as well as the re-enforcement of gender-based inequalities and discriminatory practices.

As noted earlier, in current practice, women are seen and portrayed as helpless victims; their capacities, knowledge, and skills in each stage of the disaster cycle are not recognized; women are also absent from any formal decision-making positions in emergency and recovery planning.

Common inadequacies in policy and practice of disaster management and overall development are discussed below:

> The absence of sufficient analysis of communities from a gender perspective, results in the invisibility or stereotyping of women.

The affected/displaced/beneficiaries are assumed to be families or individuals. There is no acknowledgment of their gender-based specificities. At the same time, there is a complete lack of gender sensitivity in viewing the household unit.

[1] Waheed A, Sheikh R, 'Gender and Livelihood Options for Disaster Risk Reduction, Tharparkar, Pakistan', Case Study carried out for ITDG South Asia, Journalists Resource Centre, Islamabad, 2000, p 37

[2] Chakravarti S, 'Hope in Hell' *India Today* February 26 2001 in 'We want work: Rural Women in the Gujarat Drought and Earthquake', Report based on a Quick Response Grant from The Natural Hazards Research and Applications Centre and the National Science Foundation, 2001, p 4 www.colorado.edu/hazards/qr/qr135/qr135.html

[3] ITDG South Asia, Research studies on 'Gender issues in Disasters' carried out for the regional project 'Livelihood Options for Disaster Risk Reduction in South Asia', ITDG South Asia, 2001 Enarson E, 'We want work', www.colorado.edu/hazards/qr/qr135/qr135.html Kafi SA, *Disaster and Destitute Women, twelve case studies*, Bangladesh Development Partnership Centre, Dhaka, 1992

- In the Tharparkar district, married women have to cover their faces and live in seclusion from men, which results (aside from various other disadvantages) in their inability to access relief goods directly.[1]

- Media coverage of the Gujarat earthquake showcased images of grieving and exhausted women, thereby providing visibility to women only as victims.[2]

Development concepts that view women as 'homemakers' or 'helpmates' and men as 'male earners, male breadwinners', 'male heads of households', 'the farmer and his wife' provide false pictures with regard to gender identities and gender relations. Consequently, disaster management interventions may consider the male as the 'head of the household' in relief distribution, and thereby, ignore the specific needs of women (and children) while excluding women's participation.

These concepts also give false notions about women's economic role in general, and in times of crisis in particular. For instance, studies carried out in South Asia have shown that:

- Women find it extremely difficult to find paid work in disaster situations.[3] In Chitwan, Nepal, after the floods, women and girls were desperately looking for work opportunities. In Bangladesh after the 1999 cyclone, affected women from salt-pan areas were not getting any work opportunities. Difficulties in finding paid work is reported from drought and earthquake hit Gujarat, India.

- In Surendranagar, Gujarat, India, the urgent need to earn an income is made more difficult by the routine tasks filling women's lives – caring for children, cleaning the house, courtyard, and cattle shed, attending to the ill and elderly, gathering fuel wood, water and fodder, and tending to cattle, water buffaloes, goats, chicken and other livestock.[4]

- In many drought-impacted households across Gujarat, India, women's incomes are the sole support of the family, as declining agricultural production force men out of their home-villages in search of wage labour.[5]

Insensitivity towards gender-related vulnerabilities and constraints result in denying women's needs.

The lack of sanitation facilities in the Chitwan resettlement areas poses immense problems for women and girls, due to women's need for privacy when defecating. Consequently, they have to wake early in the mornings, or wait till darkness to do so.

Lack of women medical officers in the Tharparkar district during the drought resulted in sick women not being able to get medical attention.

Disaster relief distribution in Bangladesh is almost exclusively male-dominated. All the top-level planners and decision-makers in disaster management are men, as are almost all relief workers and health workers, resulting in women's needs being overlooked.[6]

After the 1991 cyclone, families headed by women had little or no chance of getting relief. Only men had access to the relief goods distributed, while women were unwilling to come forward due to their social/cultural status - as only very poor women queue for relief goods.[7]

[4] Enarson E, 'We want work: Rural Women in the Gujarat Drought and Earthquake', Report based on a Quick Response Grant from The Natural Hazards Research and Applications Centre and the National Science Foundation, 2001, p 10 www.colorado.edu/hazards/qr/qr135/qr135.html

[5] Op. cit. p 9

[6] Masood M, Wahra GN, 'Bangladesh: learning to deal with disasters' in Fernando V, Fernando P, *South Asian Women: Facing Disasters, Securing Life*, eds, ITDG Sri Lanka for Duryog Nivaran, Colombo, 1998, p 32

[7] Bangladesh Centre for Advanced Studies, 'Cyclone 91, An Environmental and Perceptional Study', Dhaka, 1991

> Inability to provide adequate levels of security for women in emergency management efforts results in increasing gender-specific risks.

[8] Kafi SA, *Disaster and Destitute Women,* Bangladesh Development Partnership Centre, Dhaka, 1992 and Rahman MA, Rowshan D, 'Gender and Natural Disasters in Bangladesh: Strategies to Reduce Vulnerability of Women', Report on the South Asia Consultation on Natural Disaster Management, Rajapakse S, Bhagbanprakash eds, Ministry of Youth Affairs and Sports, Government of India, Commonwealth Youth Programme: Asia Centre, Bhuwaneswar, India, Nov 29-Dec 3, 2001

Security is a crucial issue for women in crisis situations. There are many reported[8] instances of women refusing to go to cyclone shelters for safety; and flood victims not going to the assigned safe shelters, when sufficient security and privacy were not ensured. Women fear sexual abuse, as well as the social consequences of being exposed to non-family males.

During floods in Multan, Pakistan, the marooned people refused to shift to the relief camps set up by the authorities. In probing the matter, it was found that not a single camp offered security to women (the wives and daughters of the families), or enclosures, where women could sleep separately in the nights. Also, the camps lacked toilet facilities, making it difficult for women to maintain their privacy amongst thousands of other marooned people.[9]

> The ignorance of the differences between the various categories of men/women results in the denial of specific needs.

Commenting on the Chitwan rehabilitation programme after floods, Kamala, a flood-affected woman, highlighted that the programme did not care much for babies, young children, pregnant women, and old men.

After the Gujarat earthquake, it was observed that young girls and elderly widows were among the groups who require specific assistance. Widows in particular, were a group who greatly needed paid work, and assistance to repair their homes.

[9] Klasra R, 'Dawn', Lahore, 15.9.97

After the floods, in Chitwan, as families disintegrated during resettlement, old people were alienated - making them insecure. Consequently, there were gaps within families for care, adequate food, and access to services.

> Lack of sensitivity towards alleviating women's roles and responsibilities results in worsening the workload of women.

For instance, the provision of dry rations and uncooked food in the immediate aftermath of a disaster will not alleviate women's situation - as they are the ones who would be expected to find water/collect fuel, and construct fireplaces, to prepare the meals for everyone. Whereas, the provision of cooked food in the immediate aftermath of a disaster will lessen women's workload considerably, and relieve stress.

The low status accorded to women's capabilities and the denial of their voices, knowledge, and experiences, result in the marginalization of women.

Women are not consulted in decision-making; nor are there participatory approaches to disaster management, even though women may have been active in disaster preparedness and possess the capabilities to suggest long-term solutions.

- There is hardly any evidence of women representatives either in disaster planning processes, or in assisting with emergency management and rehabilitation activities (including medical personnel).[10]

- In Kamra, Punjab, Pakistan, it was the women who proposed the most appropriate suggestions for early warning and long-term preparedness measures, when planning a community-based flood preparedness programme.[11]

[10] ITDG South Asia, Research studies on 'Gender issues in Disasters' carried out for the regional project 'Livelihood Options for Disaster Risk Reduction in South Asia', ITDG South Asia, 2001
Masood M, Wahra GN, 'Bangladesh: learning to deal with disasters' in Fernando V, Fernando P, *South Asian Women: Facing Disasters, securing Life*, eds, ITDG Sri Lanka for Duryog Nivaran, Colombo, 1998, p 32

[11] ITDG South Asia, 'Community based flood preparedness initiative in Kamra, Pakistan, a case study' ITDG, 2002

Lack of awareness about legal and cultural discrimination against women results in the denial of their rights.

[12] Waheed A, Sheikh R, 'Gender and Livelihood Options for Disaster Risk Reduction, Tharparkar, Pakistan', Case Study carried out for ITDG South Asia, Journalists Resource Centre, Islamabad, 2000, p 14

[13] Kafi SA, *Disaster and Destitute Women, twelve case studies*, Bangladesh Development Partnership Centre, Dhaka, 1992 ITDG South Asia, Research studies on 'Gender issues in Disasters' carried out for the regional project 'Livelihood Options for Disaster Risk Reduction in South Asia', ITDG South Asia, 2001

[14] 1 decimal equals 40 sq. metres

[15] Kafi SA, *Disaster and Destitute Women, twelve case studies*, Bangladesh Development Partnership Centre, Dhaka, 1992, p 63

- In Tharparkar women did not have access to the first step of becoming a citizen (by obtaining an identity card), due to cultural constraints on mobility and association with men, the majority of women, specially young girls, find it difficult to obtain National Identity Cards (NIC).[12]

- Studies on Bangladesh record incidents when women were not allowed to register as refugees because there was no man in the family. Female heads of households were not given shelter and food assistance.[13]

- Shamima, from Saturia, Bangladesh, narrates that after the floods killed her husband and son, her brothers-in-law occupied by force, almost all the plots owned by her husband, leaving her daughters and herself, just eight decimals[14]. She was also refused possession of the croplands her husband owned based on the plea that the Shariah law prohibited her right to land since her husband did not leave behind a son.[15]

- In Chitwan, Nepal, and in Gujarat, India, more girl children dropped out of school due to the financial difficulties (of their families), and additional domestic responsibilities, after their mothers went in search of work following floods and earthquake.[16]

Women who loose their productive resources and assets during disasters, find it extremely difficult to restore them, in the absence of any support structures during rehabilitation. Loss of assets has a direct implication on women's status in the family and in society.

- It is reported that in Chitwan, Nepal, women who lost their valuable assets in the 1993 floods were still unable to restore them - even 7 years later, when the study was conducted.

Other key gaps observed in the current disaster management practices are as follows:

- **Lack of documentation on various aspects of disaster, a lack of clear sources for information, and a lack of clearing houses resulting in false policy assumptions.**

- **Lack of consistent roles/strategies, integrated coordination in/between public/private agencies resulting in gaps and duplication.**

- **Lack of media attention on the ways and means through which affected peoples cope with situations of crisis (the media usually tends to be selective and sensationalist; focusing only on sites of misery) resulting in the underestimation of people and their contributions.**

- **Lack of awareness with regard to the political aspects of disaster/relief resulting in superficial assistance and sensationalism.**

[16] ITDG South Asia, Research studies on 'Gender issues in Disasters' carried out for the regional project 'Livelihood Options for Disaster Risk Reduction in South Asia', ITDG South Asia, 2001
Enarson E, 'We want work: Rural Women in the Gujarat Drought and Earthquake', Report based on a Quick Response Grant from The Natural Hazards Research and Applications Centre and the National Science Foundation, 2001, www.colorado.edu/hazards/qr/qr135/qr135.html

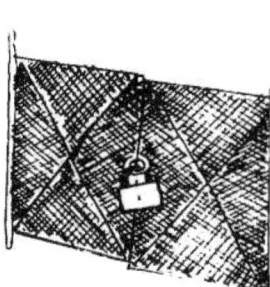

Agendas for change: policy and practice

To make development and disaster management holistic and sustainable, it is necessary to have policies that:

- Transfer the focus from emergency management to risk management (Chapter 2 deals with this paradigm shift).
- Ensure that structures, systems, and mechanisms are conducive to risk management, and that implementing officials are aware of these concepts.
- Make disaster risk management an integral part of development planning.
- Ensure that relief distribution is not arbitrary and unplanned, but integrated into long-term disaster preparedness and development.
- Reduce relief expenditure by investing more on disaster risk reduction, and on building communities that can resist disasters.
- Ensure that planning and implementation of development and disaster management are gender-sensitive to all stages of the disaster cycle.

> - Ensure the recognition of women's practical gender needs in emergencies (they are denied work opportunities, cooked food, secure shelter, water, fuel etc.)
>
> - Ensure the recognition of women's strategic gender needs (personal security, freedom from domestic violence, educational opportunities, legal rights etc.)

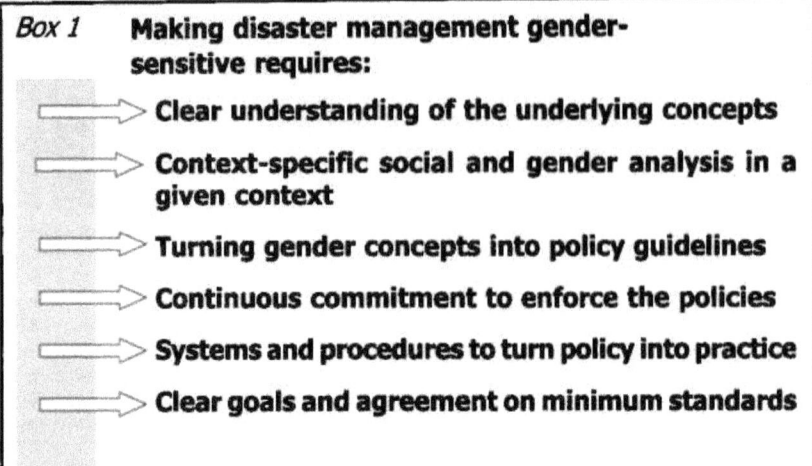

Box 1 **Making disaster management gender-sensitive requires:**
- Clear understanding of the underlying concepts
- Context-specific social and gender analysis in a given context
- Turning gender concepts into policy guidelines
- Continuous commitment to enforce the policies
- Systems and procedures to turn policy into practice
- Clear goals and agreement on minimum standards

- Understand specific gender concerns in disaster mitigation planning, and take measures to accommodate them.
- Recognize women's value/capacity as a resource in both risk reduction and in disaster management, and reject the view of women as helpless victims and liabilities.
- Provide women with the necessary information and skills so as to enable them; and reduce their overall degree of risk and vulnerability.
- Include women in all stages of disaster management plans.

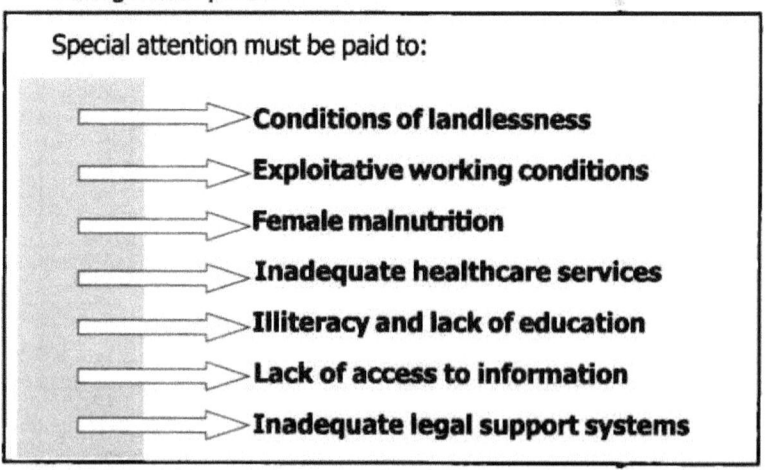

Special attention must be paid to:
- Conditions of landlessness
- Exploitative working conditions
- Female malnutrition
- Inadequate healthcare services
- Illiteracy and lack of education
- Lack of access to information
- Inadequate legal support systems

- Pay special attention to extremely vulnerable groups such as the following:

Highly vulnerable women[1]
⇨ Poor or low-income women
⇨ Refugee women and the homeless
⇨ Elderly women
⇨ Women with cognitive or physical disabilities
⇨ Women heading households
⇨ Widows and frail elderly women
⇨ Indigenous women
⇨ Recent migrants
⇨ Women with language barriers
⇨ Women in subordinated cultural groups
⇨ Socially isolated women
⇨ Caregivers with numerous dependants
⇨ Women in shelters/homeless women
⇨ Women subject to assault or abuse
⇨ Women living alone
⇨ Chronically ill women
⇨ Undocumented women
⇨ Malnourished women and girls

[1] Enarson E, Gender and Natural Disasters, Working paper 1, InFocus Programme on Crisis Response and Reconstruction, Recovery and Reconstruction Department, ILO, Geneva, 2000, p 6

- Ensure that reconstruction objectives move beyond restoring 'normalcy'; towards achieving an improved state for affected communities.

- Pay special attention to people's livelihoods in both the short and long term by introducing new options if necessary.

- Establish inclusive processes of monitoring and evaluation by sharing information with all stakeholders - whether at the level of policy formulation or at the level of implementation; with practitioners and community members, and in particular, representatives of women's groups who are knowledgeable and in close contact with the communities.

Agenda for policy makers

The following table is a guideline for policy-makers. It outlines key policy requirements for ensuring that disaster mitigation is linked to overall development planning, and that it is gender-sensitive. Examples of good practice have been provided. These are not exhaustive, and policy-makers need to take care that policies are relevant/appropriate for location, and other specificities.

Policy requirement	Good practice
1. Linking disaster management to development	
Make disaster preparedness/ management action plans mandatory in all disaster-prone areas (as an integral part of development planning).	Development plans in all hazard-prone areas to include a risk analysis that is gender disaggregated. Development plans in all hazard-prone areas to identify gender-based vulnerabilities and capacities.
Introduce policies that demand the analysis of gender disaggregated disaster risks at the level of districts and lower administrative units. This analysis would serve as the core document for the preparation of development/disaster management plans for each location.	Gender disaggregated disaster risks and needs analysis of the disaster prone areas are conducted and updated.

Policy requirement	Good practice
Provide policy support to realistically generate the required resources for district or lower level risk analysis.	Availability of long-term finances for research and technical support with skilled personnel.
Ensure that human settlement schemes are not situated in disaster prone areas.	**Resettlement in safe/viable areas** with incentives such as adequate infra-structure facilities, livelihood options, credit schemes etc.

2. Gender sensitizing disaster management

Policy requirement	Good practice
Facilitate research on gender issues in regard to disasters and in risk/hazard analysis.	Examination of the differential impact of disasters on men/women in relation to varying occupational groups, family size, caste, educational status, land ownership, age and other variables.
	Analysis of the economic and neo-economic activities performed by men and women in managing the family during 'normal' times, during disasters, and in post-disaster situations.
	More research inputs into men's gender issues such as disaster-related changes in the division of labour, men's paid and unpaid work, men as caregivers in disasters, interpersonal violence and other topics.

Policy requirement	Good practice
Target all key officials in training and awareness-building on the following: • Concepts and practices of preparedness-based disaster management • Gender-disaggregated disaster risk and needs analysis	Training and consciousness-raising on gender issues in disasters for policy makers, practitioners and communities. Challenging of gender stereotypes (for example, women as passive victims and men as invulnerable heroes). Formal gender-sensitization training to all related authorities. Informal gender training to communities. Consciousness-raising workshops/talks on gender issues at all levels. Provision of gender-sensitization training for the media to encourage accurate reporting.
Ensure policy commitment to equitable gender representation in all aspects of disaster management and development planning.	Special mechanisms and forums to ensure women's participation. Community-based, participatory methods to ensure involvement of disaster-affected women and men in all aspects of disaster management. Inclusion of women and men in decision-making forums. Inputs towards training more women as scientists, weather technicians, environmental planners etc.

Policy requirement	Good practice
Formulate policies to support the access to, and usage of, accurate statistics and gender-disaggregated data in order to provide realistic representations of disasters, and disaster-risk situations.	Gender-disaggregated data banks and regular updates. Designing and conducting of research projects based on gender-disaggregated statistics to identify the gender implications in disaster management. Making statistics/research outcomes available to those involved in disaster management, development planning, and the media. Ensuring that there are no preconceptions with regard to women (vis a vis their occupations, characteristics, etc.) in data collection.
Use gender analysis/gender sensitive methodologies in the collection of information for policy-making and implementation. (Continued on next page)	Recognition of changing family structures due to in/out migration. Accounting for the needs of widows, single parent families, pregnant and nursing women and old women. Accounting for the needs of old men, single-parent families and sick or elderly men. Recognition of the concerns of orphans, babies and children. Recognition of female heads-of-households in single-parent families.

Policy requirement	Good practice
(Continued)	Recognition of the concept of joint heads-of-households in dual-parent families. Use of appropriate and sensitive techniques to evaluate the situation of both men and women. Use of innovative methods of data collection to reflect gender interests and needs.
Adopt policies, which take into consideration specific vulnerabilities of women based on such intersects as class, race, caste, age, political affiliation etc.	Development and disaster management plans which are sensitive, and which address these vulnerabilities. Ensuring that financial resources are available to attend to specific vulnerabilities.
Formulate policies facilitating the recognition of the specific skills of women in preparedness and emergency management plans.	Practices that take into account invisible human resources and skills in disaster preparedness and management (such as women's specific skills of ecological preservation and other knowledge).
Make provision for acknowledging and valuing women's contributions and capacities in crisis situations.	Affirmation and articulation of women's contributions in needs analysis documentation. Facilitation and incorporation of such contributions in daily programs. Encouragement/incorporation of such contributions at decision-making levels within the communities.

Policy requirement	Good practice
Adopt policies that support the preservation and revival of women's livelihoods through special assistance.	Reconstruction work that takes into account the gendered division of labour in the home, in agriculture, in formal employment and the informal sector.
Formulate policies supporting alternative income generation for women and men within the disaster cycle as an integral part of development planning.	Opportunities (income generating/ training etc.) to both men and women in all stages of disaster cycle. (Women's access should not be constrained by gender stereotyping of occupations/ roles). Training in both old and new skills. Special assistance for heritage-based/traditional skills. Value-addition or innovative inputs to market women's skills and products. Easy access to distribution systems/ markets. Special attention to creating employment opportunities for displaced women, widows and women-household heads.

Policy requirement	Good practice
3. Disaster prevention	
Adopt policies, which encourage development schemes with inbuilt disaster prevention practices.	For example: Proper drainage channels, Flora replacement practices, Reforestation, Other environmental friendly practices, Housing designs/materials that can withstand earthquakes. Shelters and safe houses for cyclone flood affecting areas.
Encourage the incorporation of disaster prevention schemes/practices into the daily lives (of both men and women).	These include community-based local schemes by the government and other development and disaster management players that include: Rainwater harvesting/storage, Soil conservation, Locally appropriate early warning systems, Earthquake/landslide/flood resistant house- building techniques.

Policy requirement	Good practice
Facilitate the establishment of early warning systems with the involvement of all community groups.	Provide: Emergency codes/disaster-preparedness training to all sections (men/women/children) of the communities at risk. Family-based disaster management plans centred on practical steps to be taken before, during and after a crisis. Community-based disaster plans centred on the practical steps to be taken before, during and after a crisis.
Formulate policies that recognize and promote indigenous methods/systems of early warning.	Recognition and use of informal community networks of women. Recognition and use of indigenous methods such as observations of warning signals in nature. Recognition and use of endemic means of communication such as drumming. Taking into account references to "omens" and "superstitions". Recognition of the collective memory and socially-transmitted lessons learned through generations.

Policy requirement	Good practice
Incorporate new and technologically advanced methods/systems of early warning of disaster that can be accessed by both men/women and all groups (including illiterate people, groups in geographical isolation, etc.)	Use of appropriate media channels to reach men and women at risk. Use of other informal equipment such as sirens and sound systems.

4. Action during and after disasters

Policy requirement	Good practice
Adopt inclusive methods of registering people for assistance (that do not exclude vulnerable groups).	Ensuring that gender-sensitive data and strategies are available with regard to the distribution of food and other relief (that does not exclude any person). Ensuring that specific vulnerabilities are taken care of in the provision of assistance. Ensuring that people are not excluded/or over-enumerated in statistical analysis.
Establish simple and effective safety systems/disaster shelters that are gender sensitive.	Ensuring that there are separate areas in shelters, and privacy for women in common shelters and camps. Ensuring that there is physical protection and security for women (especially from sexual harassment).

Policy requirement	Good practice
Establish simple and effective gender-sensitive schemes to access relief, compensation etc.	Gender-sensitive methods of registration. Availability of female relief workers whom affected women can approach. uncomplicated methods of establishing proof of identity. Operation of decentralized schemes. Recognition of female-headed households/joint heads-of-households. Recognition of orphans and single people.
Make provision for trauma counselling - especially for women, as the psychological effects of disasters on women are greater due to their particular roles, responsibilities, and gender-specific vulnerabilities.	Provision of medical and psychiatric assistance/care/counselling. The availability of female health workers. Adequate training for health workers in techniques to care for women with specific vulnerabilities/psychological conditions. Medical centres/sick rooms in camp situations. Shelters for victims of violence. Provision of both allopathic and indigenous medications.

Policy requirement	Good practice
Make provision to arrest domestic violence in crisis situations.	Interventions by community support groups and the authorities to address increased incidence of domestic violence and increased incidence of gambling/drinking among men during and after disasters. Shelters to protect the vulnerable from violence. Easy and immediate access to the police, medical care, counselling services, and legal aid.
	Transmission of emergency services / bulletins via appropriate communication channels to the vulnerable. Methods to incite communal condemnation of domestic violence.
	Networking with women's support groups. Establishment of neighbourhood / camp watches. Alternative pleasure / leisure activities.

Policy requirement	Good practice
Formulate policies to deal with women's practical concerns: Take into account that women have added responsibilities during disasters.	Childcare facilities. Care of the sick and elderly. Access to fuel, water, fodder. Cooking and washing facilities. Basic resources to set-up house.
Ensure that women's concerns with regard to biology, gender, society, culture etc. are taken into account when dispensing aid.	Accessible and gender-sensitive sanitary facilities. Safety and privacy in location of medical centres, kitchens, toilets etc. Specific provisions for pregnant, lactating, menstruating women.
Ensure the access and mobility of women in camp circumstances.	Safe transport facilities for girl children travelling to schools. Proper lighting systems. Female accompaniment in situations of cultural seclusion. Communication facilities (post/telephone) that are affordable and accessible.
Make provisions to ensure women's strategic concerns (Continued on next page)	Girl children do not drop out of school in order to assist their families. Alternative education centres for girls.

Policy requirement	Good practice
(Continued)	Incentives to send girls to schools or penalties for not sending girls to schools. Facilitation of family support systems during emergencies and re-settlement.
Ensure the elimination of gender stereotyping in reconstruction planning and in providing training and other opportunities.	Women's productive and leadership roles are reflected in relief and re-construction plans. Programmes to enhance women's capacities and knowledge in disaster management.
Formulate policies to account for and to enhance women's economic role in the family and society.	Provisions to ensure the immediate assembling and stabilizing of gender sensitive economic support systems such as transport, access to markets, credit schemes, inputs towards advertising and distribution channels etc.
Establish easily accessible information banks, and methods of information dissemination on disaster management and support programs of State/NGO/INGO/community levels which target women.	Targeted radio programs for women during emergencies and reconstruction. Making provision for illiterate groups. Well-publicized and accessible information centres handled by both men and women.

Policy requirement	Good practice
5. Financial considerations	
Make provisions to integrate gender concerns into budgeting and action plans.	Ensuring that Needs Assessment Analysis and gender-disaggregated databases are updated regularly. Ensuring cadres for female professionals (in the areas of planning, emergency management, health care etc.) Alternative employment opportunities for affected women. Training in the areas where women contribute to livelihoods (livestock management and agriculture, processing, etc.) Training in alternative methods of farming and income generation.
Policies to follow up with financial allocations for development based on disaster preparedness during the post-disaster period.	Ensuring adequate provisions to continue with assistance even after the urgency of the situation ceases. Long-term livelihood options that are linked with the development plans for the area. Long-term resettlement options that link housing and infrastructure plans in re-settlement areas.

Policy requirement	Good practice
Account for disaster and related gender concerns in the government annual budget as a recurring allocation.	Research on priority areas of gender issues in disasters. Targets to realize gender-sensitive programmes on an annual basis. Promoting of targeted programmes, assessments, and sharing of good practices.
6. Accountability	
Put in place mechanisms to account for interventions in development, disaster preparedness/ mitigation, funding allocations etc. at governmental, organisational / community levels.	Checklists for assessing accountability at key stages of interventions. Use of the publicity interventions made by politicians during disasters for purposes of accountability. Making it mandatory to incorporate risk analysis of disaster preparedness and prevention measures in development plans. Introducing and publicizing mechanisms of accountability. Appointing teams/forums comprising of key stakeholders to ensure accountability.

Policy requirement	Good practice
Identify and eliminate power politics in relief operations as well as other power dynamics, and corruption.	Mechanisms to address: • Elitism, • Discrimination due to caste concerns, • Favouritism due to party politics, • Bias towards existing NGO target groups, • Male domination. Publicizing of financial assessments of crisis management and reconstruction.
Formulate policies to lessen the huge costs of relief operations and to emphasize cost-effectiveness of disaster risk management.	Relief and rehabilitation plans that are preparedness-based, and that have long-term sustainable goals. Development plans for disaster prone areas that target gradual reduction of the relief funds. Relief distribution is linked with long-term preparedness (for example, drought relief is invested in paying recipients to rehabilitate/construct water harvesting structures). Publicity to effective programmes which have resulted in the reduction of relief.

Policy requirement	Good practice
Regional bodies to make/agree on policies for inter- country accountability on disaster issues.	Regional forums set criteria, guidelines, indicators, for accountability on inter-country disaster issues. Regional policy papers that reflect gender issues. Agendas of regional bodies that include issues of gender-sensitive disaster management.
7. Monitoring and evaluation	
Discontinue policies that assess emergencies in isolation (from development concerns).	Technical capacity and financial resources needed to conduct assessments. Plans that contain properly placed monitoring and evaluation requirements, and appropriate indicators. Making available appropriate indicators to different groups when conducting assessments.
Ensure that provisions for evaluating the successes and failures of programmes are made available.	Making sure that gender-balanced community participation is taken into account in assessments. Making sure that the media highlights gender-balanced community participation.

Policy requirement	Good practice
Ensure that provisions for evaluating the successes and failures of programmes are made available.	Making sure that poor programmes are not repeated. Taking lessons from failed programmes as a key factor in planning new programmes. Publicizing of successful programmes through the media. Use of successful programmes as examples for capacity building of disaster management officials.
8. Media	
Put in provisions to ensure consistent and sensitive media coverage of disasters.	Elimination of gender stereotypes in reporting. Paying attention and highlighting the diverse gender interests/needs/capacities. Elimination of sensationalism in news coverage. Giving sensitive analysis of disaster situations.
Mobilize the media for the purposes of: • Information dissemination and consciousness-raising • Provision of emergency assistance • Allocation of funding • Accountability	Sensitizing programmes for media groups on disaster and development issues. Regular write-ups, TV programmes on disaster preparedness and development issues. Critical media coverage focussing on issues such as accountability, wastage of resources, insensitive interventions etc.

Policy requirement	Good practice
9. Advocacy	
Consciousness-raise amongst funders, private sector, investors and politicians on issues of gender sensitive disaster management and development.	Higher investment and demands by funders for gender-sensitive programmes. Provisions for legal rights, property rights and fundamental rights. Ensuring that women's participation in decision-making (household/community/political) is reflected in all stages.
Call for government-level policy papers containing holistic disaster management plans.	Disaster mitigation actions, which do not compromise long-term preparedness for short-term benefits. Ensuring that needs and priorities are not compromised for political gains.
Call for regional bodies (such as SAARC) to be co-opted for dialogue and action.	Highlighting of common issues of gender sensitive disaster management across the region. Regional policy papers reflecting issues of disaster preparedness. Identifying targets to be achieved for the entire region. Putting in place mechanisms for monitoring changes/improvements at the regional level.

Policy requirement	Good practice
Policies facilitating exchange of inter-country experiences.	Networking on the best practices of government bodies and development organisations in the region. Media sensitization for regional coverage, information sharing on gender issues and related best practises. Plans that reflect comparative analysis and lessons of inter-country experiences.

Guidelines for disaster management practitioners

Practitioners of disaster management include:
- Staff members in local, district and central government handling disaster management and development
- Staff in development organizations (international and local non-governmental organizations, community organizations)
- Organizations and officials of GOs and NGOs whose primary mandate is disaster management

Practitioners' roles involve planning, implementation and monitoring the main stages of the disaster cycle as noted below, as well as the overall development process.

	Planning	Implementation	Monitoring and Evaluation
Disaster Preparedness	✓	✓	✓
Emergency management	✓	✓	✓
Rehabilitation and Re-construction	✓	✓	✓
Overall development	✓	✓	✓

Phase I
Overall disaster preparedness

Action required	Good practice
Conduct area-wise assessments to understand the threats, levels of risk, the necessary levels of preparedness, required emergency management and rehabilitation approaches.	Assessments that are comprehensive, and which cover all the main development and disaster-related issues (See Boxes 1 and 2). Assessment teams that comprise members of the local community wherever possible.
In conducting assessments ensure that: • **Information collection and analysis is gender disaggregated and gender sensitive**	Checklists for assessment exercises include gender concerns. Assessment teams who are skilled with techniques on gender-sensitive information collection and analysis.
• **Gender dynamics, division of labour, gender-based roles and responsibilities are captured.**	Analysis that reflects in detail: the role of women, their contribution to household/human/social needs, their responsibilities pertaining to livelihood management and disaster management. Resource availability, women's specific knowledge, skills and capacities, and the constraints faced in performing their multiple roles, are reflected in detail in the analysis.

Action required	Good practice
• Coping mechanisms of men and women as well as joint activities are identified.	Inputs from the men and women in the local community are incorporated (to ensure accurate information on the coping mechanisms). Both the age-old and modern coping mechanisms are recorded.
• Specific gender-based vulnerabilities, and the vulnerabilities arising from other factors such as poverty, access to resources, caste, ethnicity, are taken into consideration.	Pay attention to the sub-categories of vulnerable groups within that of men and women.
• Availability and access to resources and other facilities such as health and education are assessed from a gender perspective.	Apart from the formal statistical information, other appropriate assessment techniques (such as interactions with the different community groups and stakeholders) are applied to ensure the accuracy of information.

Box 1

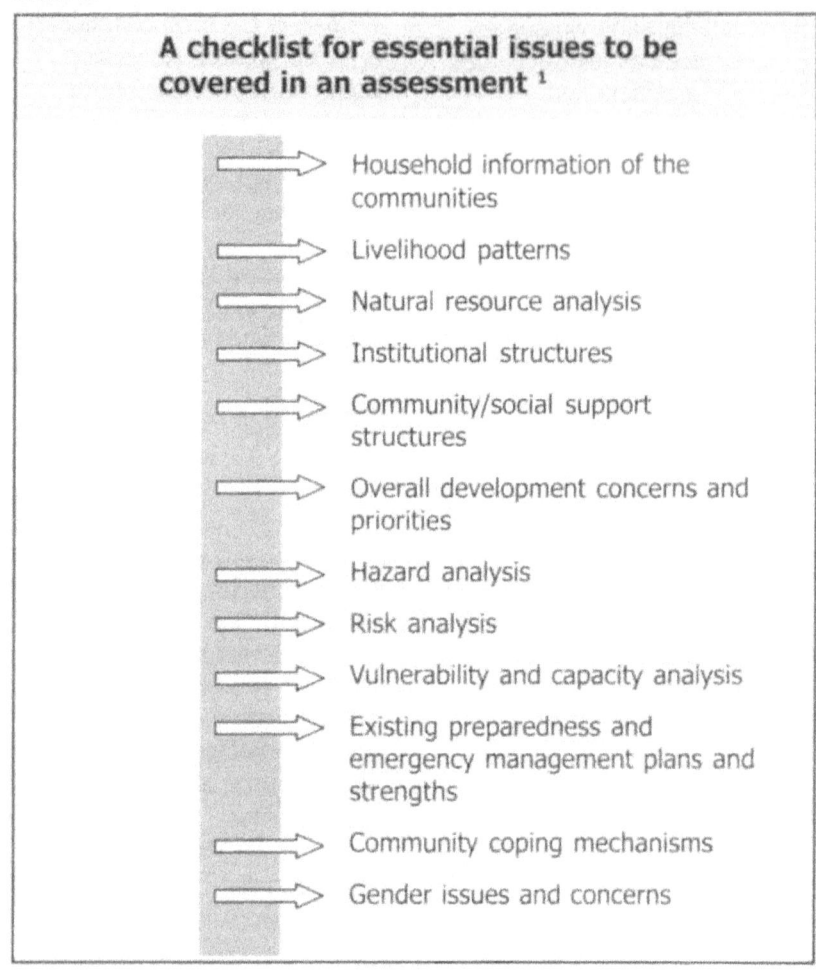

[1] Techniques and methodologies for conducting hazard, vulnerability, risk, and capacity analysis are available. Some key sources are given in the Bibliography.

Box 2

A checklist for assessment of a risk-area

⇒ Livelihood patterns and their dynamics

⇒ Resources and assets of the people in the community

⇒ Availability of water

⇒ Availability and supply of food

⇒ Availability of fodder for livestock

⇒ The formal institutional structures and accessibility

⇒ Health and education services as well as their accessibility and availability to the people

⇒ Quality and effectiveness of health services

⇒ Shelter opportunities and constraints in the area

⇒ Means of communication in the area

⇒ Employment opportunities available in the community

⇒ Available skills of the people, which can provide alternative employment

⇒ Potential and constraints to promoting those skills

⇒ Social networks which can provide safety-nets to the people during vulnerable times

⇒ Management practices of Common Property Resources

⇒ Governmental policy towards relief and development activities in the area

⇒ The work of social organizations in the communities

Action required	Good practice
Co-opt officials and professionals in development sectors, administration, infrastructure, and service-related organizations in the area when conducting the assessment.	Essential linkages with the related sectors and departments are made; and common issues/contentious areas are identified. The disaster–development linkages are clarified and maintained with relevant authorities.
Identify and develop strategies through close and regular consultations with the bodies working in the area (such as the police, the health services, the armed forces, and the government administrative agencies).	Disaster preparedness plans that are prepared in a holistic manner. Individual/institutional and joint-responsibilities of the disaster preparedness plans are clearly identified.
Based on the area assessment of the location, prepare plans for: • Disaster preparedness • Emergency management • Rehabilitation-reconstruction[2]	Plans that contain the action, the organizations/individuals responsible for each action, time frame etc. Plans that include a schedule for regular review and monitoring.

[2] Comprehensive and regularly updated area assessments can be the basis for development plans. The analysis of risk, vulnerability, natural resources, and livelihood patterns provide the fundamental basis for development planning.

Action required	Good practice
Key areas of focus in preparedness plans should include:	
1. Structural measures/ infrastructure: Drainage systems, embankments, water harvesting and storage structures, safe housing and community buildings (that can withstand earthquakes, cyclones, floods, landslides etc.), communication equipment, early warning systems, evacuation and emergency shelters, etc.	Assessments of the nature and the type of housing required for hazard-prone areas, done in collaboration with relevant Government authorities. Inclusion of emergency shelters in development proposals for hazard-prone areas. Precision-planning that ensures basic facilities (e.g. separate toilets for women and men, and measures to ensure privacy for women in emergency-shelters etc).
2. Existing institutional capacities in the area: Staff strength, level of staff knowledge, capacity for disaster response, resources available for stockpiling, storage capacity etc.	Coordination with local/district government authorities in the area.
3. Engagement of the communities in disaster preparedness. (See Box 3)	Indigenous and time-tested knowledge, informal networks and mechanisms for interaction, community-support mechanisms, ways and means of linking community knowledge and capacities with the formal institutional structures, enhancing community capacities to cope, are clearly reflected in the disaster preparedness plans.

Box 3

Engaging the Community

⇨ Identify and list the most successful strategies used by communities living in disaster-prone areas.

⇨ In consultation with the community, build on indigenous measures to develop strategies that have already been successful. The following mechanisms are recommended to integrate community participation in the planning process:

- Include both women and men from the community in the assessment, planning, and implementing of preventive measures.

- Invite women and men who are leaders in the community to strategic planning meetings and discussions, to enable a better focus on ground realities, leading to more targeted strategies.

- Identify and plan out the most useful forms of communication that can highlight and address the needs and concerns of women and men.

- Conduct workshops with children: girls and boys; women and men to devise posters, street drama, songs, plays, etc. on disaster preparedness.

- Invite schools, community-based organizations, and women's organizations to develop materials to inform the public about disaster preparedness and emergency management.

- Bring together the different village-level organizations working in the area and share the focus / mandate of these organizations.

- Make sure that community-level practitioners/ CBOs secure the involvement of both women and men in the community in local organizations.

Focus on women-specific concerns

- In consultation with the local organizations and individuals who are knowledgeable on gender and disasters, categorize:
    - The specific issues that relate to women,
    - Those that relate to children, and
    - Those that relate to the community as a whole (issues shared by women and men).

This exercise would help clarify instances where activities overlap with each other, and where activities are gender-specific.

- Identify locally appropriate, effective mechanisms to gather information/inputs from women in the area, and integrate them into the planning process.

- Pay attention to include the concerns of further vulnerable groups within the category of women (landless, widows, disabled, minority ethnic and religious groups).

- Ensure that women are not seen as 'helpless victims' by paying attention to the skills and capacities they demonstrate in livelihood and disaster management processes.

- Have separate discussions with organizations that focus on women's concerns so that the capabilities and strengths of such organizations can be enhanced through engaging them in disaster preparedness.

- Organize consultations with village-level organizations, e.g. Village Councils, Praja Sangvardana Samithi, funeral societies, Panchayet Raj, etc. that work on community issues and initiate discussions for the appointment of both women and men into leadership positions.

- Organize/mobilize special women's groups/societies in communities where it is culturally prohibited for men and women to work together.

- Formulate and ensure that literature with clear graphics/messages as well as other means of communication is available for use by women (to address the concern that women are often culturally restrained in public discussions).

- Ensure that preparedness plans contain measures to address the gender-based concerns specific to the locality, to the particular hazards prevalent in the area.

- Lay emphasis on the importance of sharing, and involving both women and men to ensure more focused action on disaster preparedness.

- Ensure that women are given space and opportunities within the planning process to apply their skills and capabilities.

Ensure that there are gender-sensitized women in staff cadre to interact with women affected/victims.

Phase II
Planning initial disaster responses/emergency management

Action required	Good practice
During each type of hazard, (floods, earthquakes, drought etc.) make note of how women and men react. **(See Box 4)**	The requirements of different community members are sensitively ascertained, and the types of rescue or relief measures are identified accordingly.
Make provisions and arrangements for accurate damage and capacity assessments after an emergency.	Skilled staff and financial provisions etc. are made available.
Be aware of the gender division of labour within and outside the households, and plan emergency responses accordingly.	Multiple roles of women, and the specificities arising from the culture, religious norms, etc, are taken in to consideration.
Be sensitive to the timing and possible location of women and men during the advent of a disaster, particularly sudden onset disasters such as floods or earthquakes.	Emergency measures take into consideration such relevant details as to whether it is a community where women are confined to the house, or whether the disaster occurred at a time when women/men are out in the fields, or a distance away from the home (collecting water or firewood, etc.).

Action required	Good practice
Take into account the special needs and vulnerabilities of women when planning evacuation, selecting/locating and managing the emergency shelters.	Procedures for evacuation and logistical arrangements within emergency shelters that are gender-sensitive. This would essentially include: • Measures to ensure the privacy of females • Measures to ensure personal security for young women in displaced conditions • Special arrangements for women with infants, pregnant mothers, old women, disabled women • Female health and relief workers are available to attend to women
Ensure that Health authorities assess the extent of diseases, injuries, loss of lives among the population, in the disaster-prone area, and attend to the needs of the affected in a culturally and gender-sensitive manner.	Female health-workers are trained and available during emergencies. Health-workers are aware of and sensitive to gender issues/ provided training on gender issues.
Take measures to attend to psychological disturbances and mental trauma through counselling.	Particular attention paid to women. Ensure the availability of female counsellors.
Develop mechanisms to build the overall gender sensitization of the authorities/stakeholders involved in rescue and emergency management.	Central Government bodies such as the Army is provided basic awareness on gender issues, and on gender-based cultural and religious specificities.

Action required	Good practice
Take measures to ensure that education and other facilities in long-term displaced situations are gender sensitive.	Girl children have access to schooling. Childcare facilities to enable women to work and engage in social activities.
Introduce measures to enhance women's' skills and capacities with awareness-building/training on emergency management, First Aid etc.	Providing information with regard to disaster-reaction procedures such as: • Locations of the emergency shelters, • Emergency meeting-points, gathering points, • Who and how to immediately contact for what problem. Including women in the emergency operations by giving responsibilities to match their skills and capabilities. Appointment of women as immediate emergency reaction coordinators.
Provide income-generation and skill-development opportunities in long-term displacement situations for both men and women.	Female trainers are available in culturally sensitive situations. Safety and logistical support (such as childcare/transport) is provided for women to engage in paid work.

Box 4

Observing gender-based responses to emergencies

For example:

⇒ What do women and men do in reaction/in relation to whom?

⇒ What are the patterns relating to who is attending to physical rescue?

⇒ Who is attending to house luggage?

⇒ Who would go in search of children first?

⇒ Who attends to the safety of the livestock?

⇒ What are the shared responses between women and men?

Action required	Good practice
Prepare gender-disaggregated assessments for relief distribution.	Sub-categories such as widows, old women, female-headed households, single women, disabled etc. are taken in to consideration.
Direct relief (or parts of the relief funds) towards future preparedness where possible.	Example: 'Food for Work' programmes coordinated with disaster-preparedness plans for the area.
Employ female relief workers.	Relief teams are gender balanced and include local women/victims in relief teams. Relief workers are made gender-aware and sensitive. Cultural concerns commonly prevalent in South Asian countries are addressed.

Action required	Good practice
Take measures to eliminate culture / religion / gender-based discrimination in registration, in compensation and relief distribution.	Relief workers are trained on ethics of humanitarian concerns. Develop mechanisms to crosscheck the accuracy of estimates and the nature of distributions across the different groups. Make it mandatory that all records are gender-disaggregated. Adhere to the minimum standards set for relief distribution (e.g. SPHERE standards[3]).
Take measures to co-opt and incorporate the skills and capacities of women in the affected community for relief planning, distribution of assistance and in other emergency management activities.	Close interaction with the affected communities during the planning process. Ensure that space is provided for the affected women to work with external teams.

[3] See 'Recommend reading' for the details

Phase III
Planning for rehabilitation/reconstruction

Action required	Good practice
After an emergency, assess the degree of damage, and record to what extent the early warnings and the preparedness measures were effective.	Assessments are not conducted in isolation, but against the earlier plans, and measures taken to identify the gaps. The gaps are assessed against the capacities of the institutions and communities. Identified gaps are addressed in forward planning.
Ensure that rehabilitation and reconstruction plans are in line and integrated into the overall development plans for the area.	Preparation of plans is coordinated with the development agencies, matched with the relevant development concerns and the plans already prepared for the area.
Make sure that rehabilitation plans aim at reaching better-prepared institutions and communities.	Avoiding the re-settlement of people in the same disaster-prone areas, or in similar positions. Aiming for higher levels of preparedness and capacities in rehabilitation planning. Addressing the gaps highlighted in the assessment plans through capacity building of the institutions and the communities.

Action required	Good practice
Incorporate the skills and capacities of the people in the area in planning and implementing the rehabilitation work.	Consultation and discussion with both women and men. Information gathered is comprehensive and locally relevant.
Deploy various techniques to get maximum community participation in rehabilitation planning.	See Box 3; Engaging the community.
Ensure that especially vulnerable groups are provided with work opportunities in the re- construction efforts.	Consulting gender-disaggregated assessment reports to identify such groups. Specific logistical support is provided to facilitate access work opportunities.

Phase IV
Monitoring and Evaluation

Action required	Good practice
Prepare monitoring and evaluation plans for each phase; disaster preparedness, emergency management, rehabilitation.	Incorporation of monitoring and evaluation plans into each phase of planning. Identifying the relevant indicators to monitor and evaluate the effectiveness of activities in each phase. Preparing plans for the regular review of monitoring and evaluation outcomes.
Select relevant indicators to monitor: • Reduced risk and vulnerability • Improved capacity (within the community and within formal institutional structures) • Mobilization of the community for collective and individual response	The selection and development of indicators that reflect gender-based vulnerabilities and capacities. Identifying specific indicators to capture the impact of disaster on specifically vulnerable groups (such as pregnant and lactating mothers, the old, female-headed households and the disabled). Selecting specific indicators to assess how women's capacities were employed and enhanced.
Build the capacities of the monitoring and evaluation teams.	The teams are gender-balanced and sensitized on gender issues.

Action required	Good practice
In all stages of monitoring and evaluation, incorporate checks and balances to ensure that the impact on the different vulnerable groups is taken into account.	Avoidance of looking at 'men' and 'women' as homogenous groups. Identifying the specific categories among men and women arising from poverty, caste, religion, ethnicity etc.
Conduct regular situation reviews of the particular hazards prevalent in the area (in anticipation of floods/ drought, earthquake etc.)	Engaging community members - both men and women - in the assessments. Setting short-term and long-term targets to monitor the levels of preparedness.
Maintain records of all the assessments for future disaster management and development planning.	Formats of records that can be easily understood. Formats and analysis are gender-disaggregated. Mechanisms that allow easy access to the records. Mandatory reference to previous assessments in the preparation of plans.
Discuss the assessment reports with the relevant groups in the community (with both men and women representatives).	Provisions are made to get the feedback from the community members.

Action required	Good practice
Make it mandatory that evaluations and damage assessments are used in planning re-construction and in future strategies of preparedness on an area basis.	Documentation is shared with relevant stakeholders. Mechanisms put in place to share information and discuss issues. Ensuring accountability in development and disaster management planning. Engaging the media in publishing key issues and in drawing the attention of the authorities.

Glossary of terms

Acid rain An industrial by-product of sulphur and nitrogen oxides emissions from burning coal and petroleum products, it should actually be called acid precipitation as it includes rain, snow, sleet, fog and any other form of acid precipitation. Found throughout the world, its heaviest concentrations are in urban areas. Among other things it harms aquatic wildlife, corrodes monuments and bridges, destroys exterior paint, kills forests and damages some agricultural soils. It makes drinking water toxic by leaching lead from pipes.

Acquired immuno-deficiency syndrome A highly infectious disease of pandemic proportions, caused by the HIV virus. A person who has the virus is a carrier and can infect others. Spread is by sexual intercourse, contaminated needles and syringes, transfusions of infected blood, and by an infected mother to her unborn child. However, spread is unlikely through daily social contact, such as shaking hands.

The Glossary is drawn from definitions developed by the following:

Asian Disaster Preparedness Center, Disaster Relief Library (http://www.disasterrelief.org/Library/Dictionary), ReliefWeb (http://www.reliefweb.org)

Asian Health Disaster Management Guidelines for Evaluation and Research in the 'Utstein-Style' http://pdm.medicine.wisc.edu/vocab.htm#natural

Forum of Environmental Journalists and ESCAP *Reporting on the Environment: A handbook for journalists*, Bangkok (1988)

Afforestation   Conversion of bare or cultivated land into forest. *See deforestation.*

Aftershocks   The long, exponentially decaying sequence of smaller earthquakes that follow a large-magnitude earthquake for months to years, exacerbating the damage.

Agricultural waste   Poultry and livestock manure or residual materials (either in liquid or solid form) generated in the production and marketing of poultry, livestock, fur-bearing animals and their products, rice straw, rice husks and other plant wastes.

Air pollution   The presence of considerable quantities of gaseous, liquid or solid contaminants in the atmosphere, that is liable to be harmful to animal, vegetable and human life.

Alarm procedure   Alerting every party concerned. Various optical and acoustical means of alarm are possible: flags, lights, sirens, radio and telephone.

Aquifer   A geological formation, which is usually composed of rock, gravel, sand or other porous material and which yields water to wells or springs. Can be polluted by introduction of pollutants through poorly capped wells, injection waste disposal and other entries below ground.

Assessment   Survey of a real or potential disaster to estimate the actual or expected damages and to make recommendations for preparedness, mitigation and relief action.

Avalanche   Sudden slide of a huge mass of snow and ice, usually carrying with it earth, rocks, trees and other debris.

B

Biological hazardous waste — Any substance of human or animal origin, excluding food wastes, which is disposed of and which could harbour or transmit pathogenic organisms. Such waste includes tissues, blood elements, excreta, secretions, bandages and related substances.

Biological warfare — The intentional spread of disease in warfare through the dispersal of infective bacteria, rickettsiae, viruses or toxins which cause diseases, such as anthrax, plague, typhoid, brucellosis. There is a UN Convention against biological weapons. Biological and chemical weapons are usually considered together (CBW). Synonyms: bacteriological warfare, biological weapon, BW.

Biomass — Any organic material that can be turned into fuel-wood; includes dry plants and organic wastes

Built environment — The buildings and lifelines (or infrastructure) of the community. *See infrastructure.*

Chemical warfare — War in which harmful chemical substances are used with the intention to kill, injure, or otherwise incapacitate humans or to destroy the environment and national economies. Chemical weapons are internationally outlawed by the 1925 Geneva Protocol. Synonyms: chemical weapon, CW.

Coral reef degradation — Caused by natural and man-made events including hurricanes, earthquakes, volcanic eruptions, disruptive invasion by marine organisms, chemical pollution, pesticide pollution, and destructive fishing methods such as dynamiting or bumping the coral bottom with fish-net weights.

Creeping disaster	A disaster of insidious onset and slow progress, such as famine, drought, desertification, health deterioration or epidemic, that does not become manifest until damage and suffering reach extensive proportions and need massive emergency response.
Cyclone/ Hurricane/ Typhoon	The terms hurricane and typhoon are regional names for a strong 'tropical cyclone.' All originate in tropical or sub-tropical waters and must spawn winds in excess of 74 miles per hour.

Hurricane - North Atlantic Ocean
Typhoon - Pacific Ocean east of the international date line
Severe tropical cyclone – Southwest Pacific Ocean, Southeast Indian Ocean
Tropical cyclone — Southwest Indian Ocean

The most intense tropical cyclone on record is Typhoon Tip in the Northwest Pacific Ocean, which on October 12, 1979 had winds gusting as high as 190 mph.

D

Damage assessment	Damage Assessment is the process of gathering timely and accurate information about a disaster area.
Deforestation	The loss of forests due to collection of fuelwood, commercial logging, shifting cultivation, grazing, road construction, ranching, mining and fire. Leads to soil erosion and flooding and endangers wildlife through habitat destruction.
Desertification	A process whereby the productivity of the land is reduced through deforestation, water logging and salinisation, chemical degradation by nutrient leaching, and rangeland mismanagement such as overgrazing, soil erosion and aridity and semi aridity.

Disaster "A situation resulting from an environmental phenomenon or armed conflict that produces stress, personal injury, physical damage, and economic disruption of great magnitude".[1]

An unforeseen and often sudden event that causes great damage, destruction and human suffering. Though often triggered by a natural hazard, disasters can have human origins. An event is classified as a disaster when it results in a serious disruption of the functioning of society, causing widespread human, material, or environmental losses which exceed the ability of the affected society to cope using only its own resources. The term is sometimes also used to describe a catastrophic situation in which the normal patterns of life (or eco-systems) have been disrupted and extraordinary, emergency interventions are required to save and preserve human lives and/or the environment. Disasters are frequently categorized according to their perceived causes and speed of impact.

[1] Cuny, FC, *Disasters and Development*, INTERTECT Press, Texas, USA, 1994

Disaster prevention The aggregate of approaches and measures to ensure that human action or natural phenomena do not cause or result in disaster or similar emergency. It implies the formulation and implementation of long range policies and programmes to eliminate or prevent the occurrence of disasters. Based on vulnerability analysis of risks, it also includes legislation and regulatory measures in the field of town planning, public works and environmental development.

Disaster management A collective term encompassing all aspects of planning for and responding to disasters, including both pre- and post-disaster activities. It may involve the management of both the risks and consequences of disasters.

Displaced person A term usually applied to people fleeing their homes because of an armed conflict, civil disturbance or natural disaster. It refers to people as long as they remain within the borders of their own country. Once they cross into another country they are defined, in most cases, as refugees.

Drought A normal, recurring feature of climate that originates from a lack of precipitation over an extended period of time, usually a season or more. Droughts can occur in virtually all climates. The precise definition depends on the region, but the definition is often determined by comparing recent precipitation to a 30-year average. In some areas, precipitation that is only 75% of a 30-year average is considered a drought.

Earth flow Mass of water-logged earth, sliding by gravity along a slope at a relatively slow speed of a few kilo meters per hour. *Compare to or see mudslide.*

Earthquake A shaking of the earth caused by a sudden movement of rock beneath its surface. An earthquake occurs on a fault, which is a thin layer of crushed rock between two blocks of rock. A fault can range in length from a few inches to thousands of miles. The San Andreas fault in California is 650 miles long and ten miles deep in places. Stresses in the earth's outer layer push the sides of the fault together. Stress builds up and the rocks slip suddenly, releasing energy in waves that travel through the rock to cause the shaking that we feel during an earthquake.

Earthquake-resistant buildings Buildings that are sited, designed and constructed in such a way that they are able to resist the ground shaking from large-magnitude earthquakes without collapsing and from moderate-magnitude earthquakes without significant loss of function and with damage that is repairable.

Ecosystem The interacting system of the biological community and its non-living environment.

Effect(s) The consequences of an event. The effects may be single, but most often are multiple and involve multiple basic functions. Any or all of the basic functions of the society affected may become impaired as part of the effects of the impact: 1) Medical services; 2) Public health; 3) Sanitation and water supplies; 4) Food; 5) Shelter/clothing; 6) Energy supply; 8) Public works; 9) Environment; 10) Logistics/transportation; 11) Security; 12) Communications; and/or 13) Economy.
In addition, the structure that provides for Coordination and Control of these functions may become impaired or inoperative. It is important to recognize that there may be primary and secondary effects of an event. Primary effects are those that are a direct result of the event. Secondary effects are those that result from the primary effects or from the responses to the event. Although described as acute, some effects may be ongoing and stretch over long periods of time (e.g. famine, drought, epidemics, complex human emergencies).

Elements at risk The population, buildings and civil engineering works, economic activities, public services and infrastructure, etc. at risk in a given area.

Emergency An extraordinary situation where there are serious and immediate threats to human life as the result of a disaster, the imminent threat of disaster, the cumulative process of neglect, civil conflict, environmental degradation and socio economic conditions. An emergency can encompass a situation in which there is clear and marked deterioration in the coping abilities of a group or community.

Erosion The loss of surface soil through the action of precipitation and wind. Leads to sedimentation and siltation of water-ways which destroy aquatic and marine habitats, make water undrinkable and clog water-dependent industrial machinery and other intake equipment.

Famine A lengthy period of time during which people experience a severe lack of food. War, poverty, drought, floods, volcanic eruptions, earthquakes and other disasters can cause famines. According to the United Nations, an estimated 20 per cent of the populations of developing countries – more than 800 million people – are food-deficient. Although worldwide food production has improved over the past few decades, famine is still a threat in many areas of the world, including sub-Saharan Africa and South Asia.

Flood Floods, especially flash floods, kill more people each year than hurricanes, tornadoes wind storms or lightning. Flood water can be deceptively strong. Fresh water moving at 4 mph (a brisk walking pace) exerts a force of about 66 pounds on each square foot of anything it encounters. Double the water speed to 8 mph and the force suddenly rises to about 264 pounds per square foot. That's enough force to punch a car or light truck off a flooded road if water reaches up to door level.

Food security Access by all people at all times to enough food for an active, healthy life. Its essential elements are availability of food and ability to acquire it. The UN Food and Agriculture Organization's definition of food security includes the following requirements: adequate supply, stable supply, and access to the supply (including adequate consumption, adequate income in relation to food prices and access to employment). Food insecurity is the lack of access to enough food. There are two kinds of food insecurity: chronic food insecurity which results in a continuously inadequate diet, and acute food insecurity which is a temporary decline in a household's access to enough food.

Gender Refers to a whole set of expectations held as to the likely behaviour, characteristics, and aptitudes men and women will have. It refers to the social meanings given to being a man or women in a given society.[2]

Gender relations Refers to the actual and perceived network of relations that occur between men and women. These involve daily, life experiences as well as notions of gender relations, which emanate from the media, religions, history, culture etc. Usually gender relations are unequal, because men have power and women do not.[3]

---

[2] Young K, Towards the theory of a Social Relation of Gender; Institute of Development Studies, Sussex, UK, 1988

[3] Wickramasinghe Maithree, Gender Identity and Gender Relations in Gender Resource Book for Teachers (ed) Mendis Some K, CENWOR 2002.

Gender needs	The different needs which arise due to the differing gender identities of men and women

> **Practical gender needs** refer to those needs that arise from women's gender roles and responsibilities such as food, water, fuel, etc. (and are usually related to immediately perceived basic needs).
>
> **Strategic gender needs** refer to those needs that arise due to women's subordinate positions in society and include such needs as equal wages, right to live free from gender based violence, legal rights etc. (and are not always visible).

Geneva Conventions	A series of international agreements that provide the legal basis for the International Red Cross and Red Crescent Movement. They reaffirm the value of human life and dignity during times of war. The first Convention protects the wounded, the sick, medical personnel and chaplains on the battlefield. The second Convention extends the protections to those at sea. The third Convention protects prisoners of war, and the fourth Convention protects civilians in enemy and occupied territories. Two additions, called Protocols, extend protection to civilian populations.
Greenhouse effect	The theory that continued burning of fossil fuels will increase concentrations of carbon dioxide in the atmosphere, thereby trapping additional heat and moisture. In time, this will raise temperature levels.
Ground water	The portion of the subsurface water which is in the zone of saturation where nearly all openings between soil particles are filled with water. The top of the zone of saturation in the ground is called the water table.

Habitat	The sum of total environmental conditions of a specific place that is occupied by an organism, a population or a community.
Hazard	A hazard is a natural or human-made phenomenon which may cause physical damage, economic losses, or threaten human life and well-being if it occurs in an area of human settlement, agricultural or industrial activity. Note, however, that in engineering, the term is used in a more specific, mathematical sense to mean the probability of the occurrence, within a specified period of time and given area, of a particular, potentially damaging phenomenon of given severity or intensity.
Hazard assessment	The process of estimating, for defined areas, the probabilities of the occurrence of potentially damaging phenomena of given magnitude within a specified period of time. Hazard assessment involves analysis of formal and informal historical records and skilled interpretation of existing topographical, geological, geomorphologic, hydrological and land-use maps.
Hazard mapping	The process of establishing geographically where and to what extent particular phenomena are likely to pose a threat to people, property, infrastructure, and economic activities. Hazard mapping represents the results of hazard assessment on a map, showing the frequency/probability of occurrence of various magnitudes or durations.
Hazardous waste	Any waste which is ignitable, corrosive, reactive or toxic and which may pose a substantial or potential hazard to human health and safety or to the environment when improperly managed (reactive refers to the ability to enter into a violent chemical reaction which may involve an explosion or fumes).

Hygienic measures	Those measures as to prevent diseases following a major disaster, because the infrastructure of the stricken area is non- or malfunctioning.
Human-made disaster	A disaster or emergency situation whose principal, direct causes are identifiable human actions, deliberate or otherwise. Apart from 'technological disasters' this mainly involves situations in which civilian populations suffer casualties, loss of property, basic services and means of livelihood as a result of war, civil strife, other conflict or policy implementation. In many cases, people are forced to leave their homes, giving rise to congregations of refugees or internally displaced persons.

Infrastructure	Structures and facilities that provide the essential functions of supply, disposal, communication, and transportation in a community. They are also called Community Lifelines.
International assistance	Assistance provided by one or more countries or international and voluntary organizations to a country in need, usually for development or for an emergency. The four main elements of assistance within the international community are:

    (a) the intergovernmental agencies—United Nations
    (b) non-governmental organizations
    (c) the Red Cross, and
    (d) bilateral agreements.

L

Landmine Any munitions placed under, on or near the ground and designed to be detonated by the presence, proximity or contact of a person or vehicle. The International Committee of the Red Cross estimates that there are 110 million active mines scattered in 64 countries. About 2 million new mines are laid each year. More than 2,000 people are killed or maimed each month in landmine explosions. Most victims are civilians. The ICRC is committed to a worldwide ban on the production, stockpiling, transfer and use of all anti-personnel mines.

Landslide A massive and more or less rapid sliding down of soil and rock, causing damage in its path. *See also avalanche, mudslide.*

Loss A range of adverse consequences impacting communities and individuals (e.g. damage, loss of economic value, loss of function, loss of natural resources, loss of ecological systems, environmental impact, health deterioration, mortality, morbidity).

M

Magnitude The overall size of the event (e.g. area of drought, number of persons potentially or actually affected, etc.). Magnitude usually expressed as a mathematical quantity.(i.e., Richter Scale for earthquakes, Safir-Simpson Scale for tropical cyclones).

Man-made disaster A disaster caused not by natural phenomena, but by man's or society's action, involuntary or voluntary, sudden or slow, directly or indirectly, with grave consequences to the population and the environment. Examples: technological disaster, toxicological disaster, desertification, environmental pollution, conflict, epidemics, fires. *See disasters, natural disasters, human-made disaster and technological disaster.*

Mitigation   A range of policy, legislative mandates, professional practices, and social adjustments that are designed to reduce or minimize the effects of earthquakes and other natural hazards on a community. Mitigation measures implemented over the last 20 years have included: 1) land use planning and management, 2 engineering codes, standards and practices, 3) control and protection works, 4) prediction, forecasts, warning, and planning, 5) recovery, reconstruction, and planning, and 6) insurance. See also disaster mitigation.

Methods of limiting damage can be as simple as placing a fuse box higher on a wall in a flood-prone area, or as costly as strengthening a building's structure to withstand an earthquake. The American Red Cross' mitigation efforts include brochures and training videos, local presentations to raise awareness of mitigation, and serving on committees and task forces that coordinate mitigation programs.

Monsoon   A weather season characterized by very heavy rainfall that affects the regions that border the Indian Ocean, especially in the Arabian Sea. The very heavy rainfall appears to result from a reversal of seasonal wind direction that blows from the southwest during one half of the year and from the northeast during the other.

Natural hazards   Natural phenomena which can occur in proximity to and pose a threat to people, structures or economic assets and may cause disasters. They are caused by biological, geological, seismic, hydrological, or meteorological conditions or processes in the natural environment (e.g. floods, severe storms, earthquakes, landslides, volcanic eruptions, wild fires, tsunamis, droughts, winter storms, coastal erosion, and space weather).

Natural disaster A sudden major upheaval of nature, causing extensive destruction, death and suffering among the stricken community, and which is not due to man's action. However, (a) some natural disasters can be of slow origin, e.g. drought, and (b) a seemingly natural disaster can be caused or aggravated by man's action, e.g. desertification through excessive land use and deforestation.

Non-governmental organization A private or international organization (as distinct from a governmental or inter-governmental organization), constituted as a single association or as a federation of various national organizations, without governmental or state ties. The most important NGOs are given consultative status with the United Nations or its specialized agencies and are active in disaster work. Acronym: NGO

Panic Acute and overwhelming sense of fear and dread, usually of sudden onset and most often self-limiting and of short duration, from a few seconds to hours, the accompanying restlessness resulting in an urge to escape. A frequent but not lasting phenomenon following disasters and major emergencies.

Parasitic diseases Infections, infestations and other disease states caused by parasites of animal origin. Some examples common in disaster situations are amoebiasis, intestinal worms, schistosomiasis, malaria, trypanosomiasis, scabies, and pediculosis.

Policy environment The forum and the process for making decisions about plans, laws, and practices to reduce unacceptable risk to people, property, and infrastructure in the community and strategic plans to implement them over time.

Pollution	Contamination of air, water, land or other natural resources that will or is likely to create a nuisance or render such resources harmful to public health or which is harmful to domestic, public, commercial, industrial, agricultural, recreational or other legitimate beneficial uses, or to livestock, wild animals, birds, fish or other life.
Post-traumatic stress syndrome	Following a period of intense stress (like in a disaster) a person may encounter short or long term psychic disorders including anxiety, insomnia, feelings of guilt, irritability, and concentration problems.
Preparedness	Measures to ensure the readiness and ability of society to forecast and take precautionary measures in advance of imminent threat, and to respond to and cope with the effects of disaster by organizing and facilitating timely and effective rescue, relief and appropriate post disaster assistance.
	Preparedness includes warning systems, evacuation, relocation of dwellings (e.g. for floods), stores of food and water, temporary shelter, energy, management strategies, disaster drills and exercises, etc. As preparedness increases, the ability of the society to absorb the event and mitigate the impact (damage) is augmented as a dependent variable of the level of preparedness.
Post-disaster assessment	The process of determining the impact of disaster or events on society, the needs for immediate, emergency measures to save and sustain the lives of survivors, and the possibilities for expediting recovery and development.

Prevention The aggregate of approaches and measures taken to ensure that human actions or natural phenomena *do not* cause or result in the occurrence of an event related to the identified or unidentified hazard. It does not mean decreasing the severity or intensity of the event. *See also disaster prevention.*

Relief Assistance in material facilities, personal needs and services given to needy persons or communities, without which they would suffer

Resiliency Pliability, flexibility, or elasticity to absorb the event. Resiliency is offered by types of construction, barriers, composition of the land, (geological base), geography, bomb shelters, location of dwelling, etc. As resiliency increases, so does the absorbing capacity of the society and/or the environment. Resilience is the inverse of vulnerability.

Risk assessment An objective scientific assessment of the chance of loss or adverse consequences when physical and social elements are exposed to potentially harmful natural and technological hazards. The endpoints or consequences depend on the hazard and include: damage, loss of economic value, loss of function, loss of natural resources, loss of ecological systems, environmental impact, deterioration of health, mortality, and morbidity. Risk assessments integrate hazard assessments with the vulnerability of the exposed elements at risk to seek reliable answers to the following questions:

1. What can happen?
2. How likely are each of the possible outcomes?
3. When the possible outcomes happen, what are the likely consequences and losses?

Reforestation The replanting of cut or bare forest. See 'afforestation'.

Refugee The fine points of definition vary under different bodies of international law, but generally, this term applies to people who have fled their country to avoid persecution or the threat of persecution on account of race, religion, nationality, membership of a particular social group or political opinion. International humanitarian law defines refugees more broadly, including displaced persons who have fled their homes during armed conflicts but have not left their homeland.

Richter Scale A scale used to measure the magnitude of an earthquake or seismic disturbance in terms of the energy dissipated. Level 2 on the Richter Scale indicates the smallest earthquake that can be felt; 4.5 is an earthquake causing slight damage, 8.5 is a very severe earthquake, likely to cause extensive damage.

Risk Risk is defined as the expected losses (lives lost, persons injured, damage to property, and disruption of economic activity or livelihood) caused by a particular phenomenon. Risk is a function of the probability of particular occurrences and the losses each would cause. Some analysts use the term to mean the probability of a disaster occurring and resulting in a particular level of loss. A societal element is said to be 'at risk' or 'vulnerable', when it is exposed to known hazards and is likely to be adversely affected by the impact of those hazards if and when they occur. The communities, structures, services, or activities concerned are described as 'elements at risk'.

Risk analysis The process of determining the nature and scale of losses and damage due to hazards which can be anticipated in particular areas during specified time periods. Evaluation of risk is the social and political judgment of various risks by the individuals and communities that face them. This involves trading off perceived risks against potential benefits and also includes balancing scientific judgments against other factors and beliefs.

Risk mapping The presentation of the results of risk assessment on a map, showing the levels of expected losses which can be anticipated in specific areas, during a particular time period, as a result of particular disaster hazards.

Run-off Water which, having fallen, flows across the surface of the ground (or a landfill or other accumulation of material), picking up materials such as soil, agricultural chemicals and other transportable material, continuing into a water course.

S

Slow disaster Usually a natural disaster with slow beginnings, which is sometimes imperceptible until the full effect is felt, as in poor crops leading to drought and famine. Synonym: creeping disaster. *See disaster, natural disaster.*

Stockpile A place or storehouse where material, medicines and other supplies needed in disaster are kept for emergency relief. Examples: UNDRO warehouse in Pisa, UNIPAC in Copenhagen.

Salinization Destruction of productive land by an increase in its salt content. Occurs frequently in over-irrigated soil when evaporation of water at the soil's surface draws up salts from underground rocks and soils, causing salts to crystallize and interfere with root growth.

Sedimentation	The accumulation of earthy matter (soil and mineral particles) washed into a river or other water body, normally by erosion, which settles on the bottom. Another use of the word is as a hazardous waste physical treatment method which separates and removes suspended particles that are heavier than the liquid in which they are present by gravitational settling.
Siltation	The same as sedimentation.
Slow-onset disasters (also called Creeping Disasters or Slow-onset Emergencies)	Situations in which the ability of people to sustain their livelihood slowly declines to a point where survival is jeopardized. Such situations are typically brought on or precipitated by ecological, economic or political conditions.
Sludge	Solid, semi-solid or liquid waste from municipal, commercial or industrial waste treatment facilities, waste water treatment plants and air pollution control facilities. In discussions of environment controls, the mud-like residue that results from the cleaning process of scrubbers or certain other devices designed to prevent solid particles from entering the environment.
Soil erosion	Movement of soil due to wind, rain and related natural forces that carry surface soil down slopes towards streams and on into rivers and eventually bays and the oceans.
Solid waste	Waste including, but not limited to, municipal, residual or hazardous waste, including solid, liquid, semi-solid or contained gaseous materials.
Species extinction	Elimination of any species of living thing as a result of habitat destruction, hunting for sport and trophies and collection and hunting for food, pleasure, research and trade. The 'incidental take' of mammals and other marine life during fishing threatens certain species.

Sudden-onset natural disasters Sudden calamities caused by natural phenomena such as earthquakes, floods, tropical storms and volcanic eruptions. They strike with little or no warning and have an immediate adverse effect on human populations, activities, and economic systems.

Sustainable development The concept of using resources in an ecologically sound manner so that they will be sustainable over the long term. Put another way, by the Executive Secretary of ESCAP, it is 'an approach to progress that meets the needs of the present without compromising the ability of future generations to meet their needs'.

Technological disasters Situations in which large numbers of people, property, infrastructure, or economic activity are directly and adversely affected by major industrial accidents, severe pollution incidents, nuclear accidents, transportation accidents, major fires, or explosions.

Technological hazard A potential threat to humans and their welfare caused by technological factors (e.g., chemical release, nuclear accident, dam failure). Earthquakes and other natural hazards can trigger technological hazards.

Tornado A violently rotating column of wind extending to the ground from the base of a thunderstorm cloud. Wind speeds can vary from 72 mph to almost 300 mph; however only about one per cent of tornadoes in the U.S. reach 200 mph wind speeds. A tornado's intensity is measured on the Fujita wind damage scale.

Toxic substances Poisonous substances known or believed to be harmful to people's health, often producing chronic, irreversible physical problems and possibly harming subsequent generations. Examples are acrylonitrile, arsenic, asbestos, benzene, beryllium, cadmium, chloroform, chromates, EDB, ethylene oxide, mercury, PCBs and many others.

Toxic waste A waste that poses a substantial present or potential hazard to human health or the environment when improperly managed. It includes wastes that are poisonous, carcinogenic, or mutagenic.

Transboundary pollution Pollution and pollutants that have been produced in one country and that have passed international boundaries through water or air to other countries, causing pollution. The effects can be mitigated only through international agreements as the damage is caused outside the boundaries of the victim country. Synonym: transfrontier pollution

Tsunami A seismic sea wave that is potentially the most catastrophic of all ocean waves. It is generated by tectonic displacement – a volcano, landslide or earthquake – of the seafloor, which in turn causes a sudden displacement of the water above and the formation of a small group of water waves having wavelength equal to the water depth (up to several thousand metres) at the point of origin. The resulting waves can be devastating to low-lying coastal areas.

United Nations The supreme intergovernmental world body established in 1945 with the purposes of 1. Maintaining international peace and security, 2. Developing friendly relations among nations, 3. Solving international problems through international cooperation, and 4. Harmonizing the actions of all nations for these common ends. The UN acts through various mechanisms, such as Specialized Agencies, e.g. WHO; Centres, e.g. Human Rights; other constituted bodies, e.g. UNHCR; committees, e.g. Disarmament; funds, e.g. UNICEF; major programmes, e.g. UNDP; peace keeping forces, e.g. UNIFIL; institutes, e.g. UNITAR, etc. UNDRO is responsible for the direction and coordination of the UN response and capability in natural and other disasters. The General Assembly has designated the 1990s as the International Decade for Natural Disaster Reduction.

Victim Casualty with sustained lesions of mechanical, chemical or nuclear nature or combinations.

Voluntary agency A non-profit, non-governmental, private association, maintained and supported by voluntary contributions. Among its activities, assistance in emergencies and disasters is notable.

Volcano A hole in the earth's crust that serves as a vent for molten rock and gases from below the earth's surface. The volcano forms from a buildup of ash and lava around the hole. There are some 1,500 volcanoes that have erupted in the past 10,000 years and are thus considered active. The deadliest eruption may have been in Tambora, Indonesia, in 1815. About 92,000 people died as a result of the eruption itself and the disease and starvation that followed.

Vulnerability The extent to which an individual, community, sub-group, structure, service, or geographic area is likely to be damaged or disrupted by the impact of a particular disaster hazard.

Vulnerability study Study and investigation of all the risks and the hazards susceptible to cause a disaster.

Vulnerability analysis The process of estimating vulnerability to potential disaster hazards of specified elements at risk. For engineering purposes, vulnerability analysis involves the analysis of theoretical and empirical data concerning the effects of particular phenomena on particular types of structures. For more general socio-economic purposes, it involves consideration of all significant elements in society, including physical, social and economic considerations (both short-and long-term), and the extent to which essential services and traditional and local coping mechanisms are able to continue functioning.

Waterlogging Soaking of agricultural land caused by a rising water-table or excessive irrigation. Compacts soil, deprives roots of oxygen and contributes to salinization.

Water pollution The introduction of substances which make water impure compared with undisturbed water. Usually this comes from soil erosion, introduction of poisonous chemicals from industries and spills and introduction of domestic sewage or industrial and agricultural wastes.

Water-table The top of the zone of saturation in the ground. *See ground water.*

Recommended reading

**Compiled by
Bilquis Tahira**

Ariyabandu, M.M.,
Defeating Disasters: Ideas for Action,
ITDG, Duryog Nivaran, 1999

1

The topics in the book ranging from vulnerability, acting against disasters, dominant and alternative perspectives, conflict, gender and disaster, priorities for action, and accountability are discussed with concise and well-reasoned argumentation supported with validating data and success stories of alternative approaches.

Ariyabandu, M.M.,
'Impact of Hazards on Women and Children: Situation in South Asia,' paper presented at "Reaching Women and Children in Disasters" Laboratory for Social and Behavioral Research Florida International University, U.S.A, ITDG, June 2000

2

The paper describes the intensity of natural disasters in South Asia, delineates the position of women and children's vulnerability to disasters as well impact of disasters on them, observations from a crisis and women's coping mechanisms, and suggests changes required.

Asian Disaster Preparedness Centre (ADPC),
A Report *on Enhancing Gender Sensitivity in Disaster Management Policies of CARE, Bangladesh,* ADPC Bangkok, Thailand,
July 1999

3

The report reviews the policies, procedures and activities of the Disaster Management Unit of CARE Bangladesh, and the gender component of the training module on disaster management, and makes recommendations for future work.

> Asian Disaster Preparedness Centre (ADPC),
> *Community Based Disaster Management:
> Trainer's Guide,* ADPC Bangkok, Thailand, 2001

4

The trainers' guide provides tools to disaster management trainers in a systematic manner on the following: the Asian disaster situation, framework for community based disaster management, community based risk assessment, gender and livelihood issues, strengthening local capacities and planning at the local level.

> Bamberger, M., (et al)
> Integrating Gender into PRS, Draft for
> Comments, World Bank, April 2001

5

The draft provides an in-depth discussion on integrating gender into Poverty Reduction Strategies and is designed to guide those involved in poverty reduction strategies (PRS) at the country level in identifying and implementing policies and programs that will benefit both men and women and maximize potential benefits for poor families.

> Bhatt, M.R., Kropac, M., and Kikani, H.,
> *Institutionalising Mitigation* DMI,
> September 2002

6

How can mitigation be institutionalized? What form does mitigation takes at local level? And how can community capacity be developed to mitigate disaster impact and risks? This annual report of the Disaster Mitigation Institute for the year 2000 to 2001 throws light on how a small, community-based and local organization in India institutionalized its mitigation activity where centrality of women in mitigation is valued and highlighted. It is an account of activities as well as a guide for similar action elsewhere.

> Bhatt, M.R., Kikani, H., and Sadhu, H., *The Victims' Voices*, DMI, February 2002

7

Many men and women talk about women's issues in a disaster relief to mitigation stages, but very few listen to what the women have to say. In October 2001 the DMI team reached out to 50 urban and rural settlements to listen to what an estimated 2300 women concerns about the performance of a food, shelter, water, and livelihood relief after the January 26, 2001 earthquake in Gujarat. Special tools were used to capture their voices from literate and illiterate women.

> CAPART (Council for Advancement of People's Action and Rural Technology), *Disaster Preparedness: A Handbook for Trainers*, CAPART, New Delhi, 1990

8

The handbook gives various ideas and suggests sessions for training in disaster preparedness.

> Carol, M., and Shahra, R., *Gender Analysis: Alternative Paradigms*, UNDP, May 1998

9

This study reviews the growing body of work on gender analysis, including recent approaches, explicating their main lines of convergence and difference and assessing the results of their incorporation into training packages and programmes.

> Centre for Policy Studies, *Impact of Disaster On Gender; A Case Study Of Flood in Nepal* for Intermediate Technology Development Group, Nepal Country Office, September 2000

10

The report analyses the roles and responsibilities of women and men in disaster situations and highlights gender related impact of disaster.

> DMI, *The DEC Response to the Earthquake in Gujarat,* DMI, Humanitarian Initiatives (HI), UK and Mango, UK, December 2001

11

This is an independent evaluation of the response of the Disaster Emergency Committee (DEC) to the earthquake in Gujarat. The DEC requested the team to focus on targeting, sheltering and financial management. In the evaluation the total response of the DEC rather than the performance of the individual members is reviewed. The report focuses on the issues arising for DEC, assessment of response against the Red-Cross Code, and lessons for the practice of disaster response. The report draws attention to the lack of serious consultation with women or gender analysis by DEC agencies.

> Enarson, E., *Gender Equality, Environmental Management, and Natural Disaster Mitigation: Report from the Online Conference Conducted by the Division for the Advancement of Women,* DAW, November 2001

12

The report highlights a gender approach to disaster management, recognizing the need to empower women and foster community involvement in all stages of disaster mitigation.

> Enarson, E. and Morrow, B.H., (eds) *The Gendered Terrain Of Disaster Through Women's Eyes,* International Hurricane Centre, Laboratory for Social and Behavioural Research, Florida International University, Miami, Florida, 1998

13

The book holds useful information for planners in the field of emergency preparedness, response and recovery.

> Enarson, E., *Working With Women At Risk: Practical Guidelines For Assessing Local Disaster Risk,* International Hurricane Centre, Florida International University, April 2002

14

This is a step-by-step guide for assessing the resources and vulnerabilities of communities "through the eyes of women." The model includes: identification of a local community women's group to take on the project; training of women from the group as community researchers; guidelines for using various research strategies to collect original data about local hazards and risk; and ideas for synthesizing and utilizing the findings.

Enarson E., *Gender and Natural Disaster* Working Paper 1, In Focus Program on Crisis Response and Reconstruction, ILO Recovery and Reconstruction Department, Geneva, September 2000

15

The Working Paper provides a valuable analysis of the gender facets of natural disasters including the gendered economic impacts in the form of: loss of assets and entitlements; increase in women's workload and care-giving functions, deterioration in working conditions, and women's rather slow recovery from economic losses. Some impacts of disasters on men are also highlighted. Also of significance is the fact that the data assembled in the document cover both developing and developed countries. Action proposals are made on how the identified critical gender aspects can be taken into account in crisis response and reconstruction.

Enarson, E., *Responding to Domestic Violence in Disaster: Guidelines for Women's Services and Disaster Practitioners*, Disaster Preparedness Resources Centre, University of British Columbia, November 1997

16

The article explains why disasters are not only powerful physical events but complex social experiences for individuals, households, and communities. Once considered the great "leveler" impacting poor and rich alike, it is obvious now that floods, hurricanes, tornadoes, chemical spills, and other environmental and technological disasters hit some social groups more than others. The poor, subordinated racial or ethnic groups, refugees and migrants, seniors, the disabled, and women are especially vulnerable to disaster losses and gender-based violence and often recover from disaster more slowly.

| Fernando, P. and Fernando, V., *South Asian Women: Facing Disasters, Securing Life*. Colombo: Duryog Nivaran Secretariat, ITDG, 1997 | 17 |

This book focuses on the interaction of gender and politics in the "management" of disasters in South Asian societies. It contains case studies from the sub continent. A video documentary under the same title is also available.

| Graham, A., *Gender Mainstreaming Guidelines for Disaster Management Programmes, A Principles Socio Economic and Gender Analysis (SEAGA) Approach*, United Nations Division for Advancement of Women, 2001 | 18 |

The paper outlines the role of socio-economic and gender analysis in addressing the need to tackle root causes of vulnerability to natural disasters.

| Gurumurthy, A., *Women's Rights And Status: Questions Of Analysis And Measurement*, UNDP, May 1998 | 19 |

The report provides a comparison of gender-related analytical/ training frameworks such as women's empowerment framework, UNDP training for gender mainstreaming and socio-economic and gender analysis (SEAGA) approach.

| IBRD/The World Bank Policy Research Report, *Engendering Development Through Gender Equality in Rights, Resources and Voice*, Washington D.C., May 2001 | 20 |

The report provides policy makers, development specialist and civil society members many valuable lessons and tools for integrating gender into development work. The evidence and analysis in the report can inform the design of effective strategies to promote equality in women and men in development.

> ILO, *Coping Strategies and Early Warning Systems of Tribal People in India in the Face of Natural Disasters*, Case Studies in Mayurbhanj, Orissa and Dungarpur, Rajasthan, India, an IFP/CRISIS INDISCO Study, ILO New Delhi, undated

21

The book provides an insight into indigenous early warning systems and coping strategies of people in disaster situations. It also focuses on the special strengths such as social capital, solidarity and ability to survive on a wider variety of nutritional resources. The coping patterns in both the communities differ somewhat, however, the wealth of knowledge in the communities and the ability for collective action is their means of survival. The report also makes recommendations on how the government can fulfil its responsibility to these communities pre, during and post disaster.

> ILO, *Crises, Women and other Gender Concerns*, Working Paper 7, In Focus Programme on Crisis Response and Reconstruction, Recovery and Reconstruction Department, Geneva, February 2002.
> Gender Guidelines for Employment and Skills Training in Conflict-Affected Countries, International Labour Organisation, Geneva, Switzerland, March 1998

22

The guidelines have been developed taking into account the special gender concerns in the conflict context in designing, implementing and evaluating skills training and employment promotion programmes. The guidelines are geared to facilitating the mainstreaming of the complex gender issues in policies and programmes in the specific context of conflict, as a tool for action. The guidelines will also contribute to advance discussion, advocacy and action at different levels.

> Hameed, K., *Gender Issues in Livelihood and Flood Disaster, Case Studies of Kamra and Kot Murad Villages*, Jhang District, Pakistan, JRC and ITDG, 2000

23

The report provides gender livelihood analysis of two villages, looks at gender vulnerabilities of disasters and provides four case studies highlighting gendered experience of disasters.

> Kottegoda, S., *Livelihood Options For Disaster Risk Reduction In South Asia: A Study Of Gender Aspects Of Communities Living With Drought And Landslides In Sri Lanka*, study undertaken for ITDG, Sri Lanka, June 2001
>
> 24

The study aims at addressing several issues regarding gender relations in disasters particularly focusing on livelihood options and direct and indirect impact on gender relations. It also presents four Sri Lankan case studies that highlight gender issues.

> Krishnaswamy, P.B., Kumar, S., Dave, M., *Gender Issues in Livelihood Options for Disaster Risk Reduction*, DMI India and ITDG, May 2001
>
> 25

The paper discusses the conceptual framework useful for gender analysis and gives an overview of gender relations, the status, impact of disasters and the role of external intervention and gives recommendations for fostering more equitable gender relations in disaster mitigation.

> Kropac, M., *Urban Development And Disaster Mitigation*, DMI, November 2002
>
> 26

Nearly two years after the January 2001 earthquake, the effect can be seen everywhere. In spite of the abundance of relief that landed in Bhuj, the government and NGOs almost bypassed the men and women in the 37 slums of Bhuj. This issue leads one to look at rebuilding after disasters as an opportunity for better urban development and focuses on importance of livelihood support and information for vulnerability reduction of men and women. This report highlights the special needs of women's economic recovery and better-targeted demands from women.

> Makwana, G., Chavda, L., *Importance Of Women's Role In Disaster Mitigation* DMI and United Nations Development Programme (UNDP), July 2001
>
> 27

The publication (in Gujarati), draws on the experience of DMI's work with women in drought, flood, cyclone and earthquake relief, and highlights the methods of increasing women's participation in relief and rehabilitation projects. It covers issue of preparedness, relief distribution, gender sensitive rehabilitation, and is illustrated with voices of women victims from the Latur earthquake.

> Anderson, M., 1994, "Understanding the disaster-development continuum: Gender Analysis is the essential tool", in *Focus on Gender*, 2/1 on Gender and Disaster Network site. <http://online.northumbria.ac.uk/geography_research/gdn/resources/bib01.html>   28

The paper effectively traces the links of disaster as a development issue.

> Rahman, M.D.S., *Disaster Management Handbook for Bangladesh*, Bangladesh Disaster Preparedness Centre and PACT Bangladesh, Dhaka, February 1993   29

A ready reference manual for people and organizations during the disaster response period.

> Munasinghe, M. and Clarke, C., *Disaster Prevention for Sustainable Development: Economic and Policy Issues*, A report from the Yokohama World Conference an Natural Disaster Reduction May, 23-27, 1994, The international Decade for Natural Disaster Reduction and the World Bank, 1994   30

The report is a concise review of the economic and policy issues relating to disaster prevention in sustainable development.

> Sequeira, N., *Risk Management: an alternative perspective in gender analysis*, UN Division for the Advancement of Women (DAW), November 2001   31

The paper presents an alternative reading of disasters arguing that the principal opportunities to factor gender considerations into risk management occur at the local level.

> Orstad, L., *Tools for Change, Emergency Management for Women,* United Nations Division for Advancement of Women, International Strategy for Disaster Reduction (ISDR), Expert Group Meeting on "Environmental management and mitigation of natural disasters: a gender perspective", at Ankara, Turkey, 6-9 September, 2001

32

The paper deals with violence against women in disaster situations and presents a strategy for women to mobilize to form groups within their communities to develop emergency plans and support each other.

> Oxfam, Humanitarian Charter and Minimum Standards in Disaster Response, The Sphere Project, Geneva, Switzerland, 2000

33

Through this Charter defined levels of service in water supply, sanitation, nutrition, food aid, shelter, site planning and health care are linked explicitly to fundamental human rights and humanitarian principles. An addendum on gender concerns has been added in 2001.

> Qaisrani, S.M., *Vulnerability Mapping of the Flood Affected Areas of Dera Ghazi Khan District,* Oxfam GB Pakistan, Islamabad, 1997

34

An assessment report focusing on the flood preparedness issues.

> Schalkwyk, J., *Exercises in Gender Mainstreaming, Gender and Development* Monograph Series # 8, Gender and Development Programme, UNDP, New York, May 2000

35

The monograph presents a group of exercises to enable gender focal points to identify gender gaps in programming and implementation. The case studies on poverty, governance, human rights, post-conflict initiatives and water resources address gender equality issues in mainstreaming programming.

| Shah, H., et al., *Making Development Gender Sensitive, A Guide for Trainers*, Ahmedabad; International Centre for Entrepreneurship and Career Development (ICECD), 1999 | 36 |

An informative guide on gender sensitization and planning, to guide development planners, workers and trainers to formulate strategies and facilitate gender equity at all levels. The manual through its wide collection of theme papers and training modules gives insight of a gender framework to be applied in different fields of the developing sector.

| Shapan, A., et al., *People's Participation NGOs and the Flood Action Plan*, Oxfam, Bangladesh, Dhaka | 37 |

The report studies the participation of grassroots communities in the Flood Action Plan.

| Rajakarunanayake, S., and Ariyabandu, M.M., *Seeing Disasters Differently: Visions and Suggestions*, A Duryog Nivaran Publication, 1999 | 38 |

The book defines disasters as a process. It goes back to show that in religious and historical texts disasters were always explained as processes.

| Thardeep Rural Development Programme, Gender Situation Analysis in Tharparkar, TRDP, Pakistan, 2002 | 39 |

The study presents the gender profile of Tharparkar; detailed statistics and analysis on the issues of mobility and access to work, property and inheritance issues, marriage rights and responsibilities, crime and punishment.

| UN/ISDR, Commission on the Status of Women, *Gender Mainstreaming in Disaster Reduction*, Geneva, March 2002 | 40 |

The paper presents deliberations on disaster impact, strategic components of disaster reduction, women as actors for change, linking gender issues in disaster reduction to sustainable development, and makes recommendations for future action.

> Vaux, T., *Disaster and Vulnerability,*
> DMI & SEWA, June 2002   41

This issue documents the Self Employed Women's Association's (SEWA) response to 2001 earthquake in Gujarat and focuses on the role of a large 530,000 strong, membership-based poor women's organisation in disaster response and recovery. The report discusses what makes women more vulnerable to disasters, and how they are able to lead some of the most difficult recovery processes.

> Wiest, R.E., Mocellin J.S.P., and Motsisi D.T.,
> *The Needs of Women in Disasters and Emergencies,* UNDP & Office of the United Nations Disaster Relief Coordinator,
> Geneva, June 1994   42

The study provides a general overview of the problems experienced by women in disasters and emergencies. It also addresses the gender bias in disaster-related research, the critical analysis of established roles of women, and the attention needed by operational agencies to the special needs of women together with dependent children.

> Zafar, A., *Proceedings of the Flood Disaster Preparedness Workshops at Multan, Muzaffargarh and Sargodha,* June-July 1994, Pattan Development Organisation, 1994   43

The report aims to strengthen flood-prone communities to be better prepared to combat disasters and minimize losses resulting from floods and highlights women's and men's issues during disasters.

Internet resources

1. For an extended list of Internet sites dealing with hazards, see http://www.colorado.edu/hazards/sites/sites.html]

2. http://www.redcross.org.uk
   (Click on "Our Work," then on "International Activities", then on "NGO Disaster Mitigation and Preparedness Project")
   For the past two years, a team funded by the Department for International Development (DFID) and managed by the British Red Cross has been researching the work of NGOs in natural disaster mitigation and preparedness. The project's findings are now available on-line from this web site. They comprise:
   1. An overview paper that summarizes the research findings.
   2. A series of short case studies for project planners, illustrating the range and nature of NGO work in this area and highlighting key issues. When it is complete, the series will contain between 15 and 20 case studies.
   3. A research study of the mitigation/preparedness work of international relief and development NGOs based in the U.K. and the factors affecting this work.
   4. Similar research studies of NGOs in Bangladesh, Nicaragua, the Philippines, and Zimbabwe.

   Questions and comments can be directed to the project's e-mail address, dmp@redcross.org.uk, or the research team leader, John Twigg, j.twigg@ucl.ac.uk.
   http://www.wa.gov/wsem

3. http://www.duryognivaran.org
   Duryog Nivaran (Sanskrit word: means disaster mitigation) advocates alternative perspective on disasters and looks into the social dimensions of natural disasters. The site contains research on the issues of livelihoods and disasters, case studies containing best practices on community based disaster risk reduction from South Asian countries, information on the publications of the network, and a photo gallery depicting various disaster situations.

4   http://geoinfo.usc.edu/gees/
    The Geotechnical Earthquake Engineering Server - a project/ service supported by the National Science Foundation - has published the "Preliminary Report of the India-US Geotechnical Earthquake Engineering Reconnaissance Team" that examined the quake in Bhuj, India.

5   http://www.nicee.org/NICEE/Gujarat/iaeemanual.htm
    http://64.177.169.147/NICEE/Gujarat/iaeemanual.htm
    In response to the catastrophic Indian earthquake, the National Information Center for Earthquake Engineering (NICEE) at the Indian Institute of Technology (IIT), Kanpur, India, has made the International Association for Earthquake Engineering (IAEE) manual "Guidelines for Earthquake Resistant Non-Engineered Construction" available via the NICEE web site. Non-engineered buildings are defined as those that are spontaneously and informally constructed using traditional techniques without the aid of an architect or engineer but that may follow a set of recommendations derived from observed behaviour of such buildings in past earthquakes and trained engineering judgment. Questions or comments about these guidelines can be directed to NICEE via e-mail: nicee@iitk.ac.in.

6   http://www.benfieldhrc.org/
    The Benfield Hazard Research Centre comprises three groups: Geological Hazards, Meteorological Hazards & Seasonal Forecasting, and Disaster Studies and Management. The Benfield HRC provides a conduit for the transfer of cutting-edge natural hazard and risk research, practice, and innovation from the academic environment to the business world and government and international agencies. Through the rapid application of new research and practice, the Centre facilitates the improvement of natural hazard and risk assessment and the reduction of exposure to natural catastrophes.

7   http://www.eeri.org
    The Earthquake Engineering Research Institute (EERI) has posted preliminary reports and photos from the EERI reconnaissance team that examined the Bhuj earthquake that devastated the state of Gujarat, India, on January 26, 2001. The site also offers observations and information about the recent Washington State earthquake and the two El Salvador earthquakes.

8 http://www.itdg.org/
ITDG – the Intermediate Technology Development Group – aims to demonstrate and advocate the sustainable use of technology to reduce poverty in developing countries.

The website contains information on ITDG's approach to strengthen the ability of poor people to use technology to cope with threats from natural disasters, environmental degradation and civil conflict by: strengthening the ways that people who live in fragile environments cope with the environmental degradation which threatens their livelihood opportunities; improving vulnerable communities' ability to prepare for, survive and rebuild homes and livelihoods after natural disasters; preventing and managing conflicts over scarce natural resources and competition for common property resources; rebuilding the livelihoods of people affected by civil war or conflict.

9 http://www.crid.or.cr/crid/Indexen.htm
The second edition of the regional disaster preparedness and management newsletter/magazine, "ISDR Informs-Latin America and the Caribbean," is now available on the Regional Disaster Information Center (CRID) web site above.

10 http://www.unisdr.org/unisdr/aboutisdr.htm
UN/ISDR website

11 http://www.anglia.ac.uk/geography/gdn

The Gender and Disaster Network is an educational project initiated by women and men interested in gender relations in disaster contexts. The network intends to document and analyze women's and men's experiences before, during, and after disaster, situating gender relations in broad political, economic, historical, and cultural context.

12 http://www.pep.bc.ca
(Click on "Tools for Change: Emergency Management for Women's Services)
or go directly to:
http://www.pep.bc.ca/management/Women_in_Disaster_Workbook.pdf
The British Columbia Provincial Emergency Program (PEP) web site (see DR #338) has made this entire workbook on disaster preparedness and response among women's services available on-line. "It Can Happen to Your Agency - Tools for Change: Emergency Management for Women's Services," prepared by the B.C. Association of Specialized Victim Assistance and Counseling Programs, focuses on how women's service agencies can prepare to meet the problems and increased demands for services that will accompany any disaster.

Gender training packages and materials

1   British Council, **Gender and Economic Reform Through Women's Eyes,** Gender Team, The British Council, Manchester, n.d.

2   Fernando, P., Appleton H., Wijethilake S., **Discovering Technologists; women's and men's work at village level**, ITDG, 2000

3   IDS, **Gender and Third World Development**, Modules 1-7, Institute of Development Studies, Sussex, n.d.

4   Kelleher, D., et al., **Building a Global Network for Gender and Organisational Change**, Proceedings of a conference on Building a Global Network for Gender Change, Toronto, 1996

5   Norem, R., **Socio-economic and Gender Analysis (SEAGA) User's Handbook: a Conceptual Approach to Development Planning, Implementation, Monitoring, and Evaluation,** for FAO and ILO, working draft, September 1996

6   Parker, A.R., **Another Point of View: A Manual on Gender Analysis Training for Grassroots Workers**, UNIFEM, Women, Ink. New York, 1993

7   Parker, A.R., I. Lozano, L. Messner, **Gender Relations Analysis: A Guide for Trainers**, Save the Children, New York, 1995

8   UNDP, **Consultation Briefing for Gender Focal Points**, New York, 3-4 February 1997

9   UNDP, **Guidance Note on Gender Mainstreaming, Senior Management Consultation on Gender Mainstreaming**, New York, 5-7 February 1997

10   UNICEF, **Gender Equality and Women's Empowerment: A UNICEF Training Package**, UNICEF, New York, 1994

11   USAID, **Gender Analysis Tool Kit**, U.S. Agency for International Development, Office of Women in Development, Washington, DC. van Staveren, I. 1995. 'Reader: Gender and Macro Economic Development,' mimeo, Oiklos, Utrecht

12   Williams, S., **The Oxfam Gender Training Manual**, an Oxfam Publication, Oxford, 1994

13   Woroniuk, B., **CIDA, WID, and Gender Equity: So What's Leading Edge**, Briefing Module, CIDA, 1995

Internet sites for gender

United Nations agencies

Women Watch (UN Internet gateway on the advancement and empowerment of women)
http://www.un.org/womenwatch/about/index.html

International Research and Training Institute for the Advancement of Women (INSTRAW)
http://www.un.org/instraw/

The United Nations Division for the Advancement of Women
http://www.un.org/womenwatch/daw/

United Nations Development Fund for Women (UNIFEM)
http://www.unifem.undp.org/

United Nations Educational, Scientific and Cultural Organization (UNESCO)
http://www.unesco.org/women/index.htm

United Nations Development Program (UNDP)-Gender in Development home page
http://www.undp.org/gender/

The Women and Habitat Programme
http://www.unchs.org/unchs/english/women/womenbody.htm

United Nations Population Fund (UNFPA)
http://www.unfpa.org/tpd/gender/aboutgenderteam.htm

United Nations Children's Fund (UNICEF)
http://www.unicef.org/ **(search facility)**

United Nations High Commissioner for Refugees (UNHCR)
http://www.unhcr.ch/issues/women/women.htm

United Nations Economic Commission for Latin America and the Caribbean:
"Gender indicators for follow-up and evaluation of the Regional Programme of Action for the Women of Latin America and the Caribbean, 1995-2001, and the Beijing Platform for Action" (English version)
http://www.eclac.org/English/research/women/indicators/genderind.htm

World Food Programme (WFP) http://www.wfp.org/genderweb/

Food and Agriculture Organization (FAO) http://www.fao.org/Gender/gender.htm
http://www.fao.org/sd/seaga/default.htm

FAO Socioeconomic and Gender Analysis Programme
http://www.fao.org/sd/WPdirect/default.htm

Women and Population section of Sustainable Development Dimensions, a service of the Sustainable Development Department (SDD) of the FAO International Fund for Agricultural Development (IFAD) http://www.ifad.org/pub/other/!brocsch.pdf

International Food Policy Research Institute (IFPRI)
http://www.cgiar.org/ifpri/themes/mp17/gender/2gender.htm

International Labour Organization (ILO) Tools for mainstreaming gender concerns)
http://www.ilo.org/public/english/region/asro/mdtmanila/training/unit5/refsrurl.htm

Multilateral organizations

Asian Development Bank (ADB) http://www.adb.org/Work/Policies/Gender_Devt/

Commonwealth Secretariat http://www.thecommonwealth.org/
http://www.thecommonwealth.org/gender (under construction)

Inter-American Development Bank (IDB) http://www.iadb.org/sds/wid/

Pan-American Health Organization (PAHO)
http://www.paho.org/english/hdp/hdwmuje.htm

World Bank on Gender http://www.worldbank.org/gender/

Bilateral organizations

U.S. Agency for International Development (USAID)
http://www.genderreach.com/default2.htm

Japan International Cooperation Agency (JICA)
http://www.jica.go.jp/E-info/E-earth/E-wid/Index.html

Non governmental organizations

Women, Ink. http://www.womenink.org/

WomenAction http://www.womenaction.org/

Self-Employed Women's Association (SEWA)
http://www.sewa.org/

ENERGIA International Network on Women & Sustainable Energy http://www.energia.org/

WIDNET http://www.focusintl.com/widnet.htm

InterAction http://www.interaction.org/caw/index.htm

Asia Foundation Global Women in Politics Program
http://www.asiafoundation.com/programs/
prog-area-wome.html

Save the Children
http://www.savethechildren.org/women.html

International Center for Research on Women (ICRW)
http://www.icrw.org/

International Forum for Rural Transport and Development
http://www.gn.apc.org/ifrtd/nletter/nleng64.htm

Centre for Population and Development Activities (CEDPA) http://www.cedpa.org/

Gender and Disaster Network - Selected Resources on Gender & Disaster [Revised 7/03]. Compiled by Elaine Enarson http://online.northumbria.ac.uk/geography_research/gdn/resources/bibliographies.html

Academia

Women and Gender in Global Perspectives Program at the University of Illinois at Urbana-Champaign
http://www.ips.uiuc.edu/wggp/

Women's Human Rights Resources at the University of Toronto Faculty of Law
http://www.law-lib.utoronto.ca/diana/

Briefings on Development and Gender (BRIDGE) at the Institute of Development Studies, University of Sussex
http://www.ids.ac.uk/bridge/index.html

Gender Dimensions in Disaster Management

A Guide for South Asia

ITDG Publication

www.ingramcontent.com/pod-product-compliance
Ingram Content Group UK Ltd.
Pitfield, Milton Keynes, MK11 3LW, UK
UKHW021846140426
5217IPUK00022B/1620